EMBRACING MATH

Cultivating a Mindset for Exploring & Learning

Deanna Pecaski McLennan

National Association for the Education of Young Children
Washington, DC

naeyc®

National Association for the Education of Young Children
1313 L Street NW, Suite 500
Washington, DC 20005-4101
202-232-8777 • 800-424-2460
NAEYC.org

NAEYC Books

Senior Director, Publishing and Professional Learning
Susan Friedman

Director, Books
Dana Battaglia

Senior Editor
Holly Bohart

Editor
Rossella Procopio

Senior Creative Design Manager
Henrique J. Siblesz

Senior Creative Design Specialist
Gillian Frank

Senior Creative Design Specialist
Charity Coleman

Publishing Business Operations Manager
Francine Markowitz

The following chapters were previously published in the specified issues of *Teaching Young Children:* Chapter 2—December 2016/January 2017; Chapter 5—February/March 2017; Chapter 6—October/November 2014; Chapter 9—April/May 2017; Chapter 10—October/November 2018; and Chapter 11—October/November 2019.

The following chapters were previously published in the specified issues of *Young Children:* Chapter 12—November 2012; Chapter 13—July 2015; and Chapter 15—March 2018.

Photo Credits

Photo illustrations by NAEYC: 57, 58, 59, and 99

Courtesy of the author: 11, 15, 19, 20, 21, 22 (both), 23, 27, 36 (top, right bottom, and left), 40 (right bottom and left), 42 (top and right middle), 46, 49, 52, 54, 55, 56, 93, and 105

Copyright © Getty Images: cover, iv, 2, 3, 4, 6, 12, 13, 14, 16, 17, 18, 28, 30, 31, 33, 34, 35, 36 (right middle), 38 (all), 40 (top and right middle), 42 (right bottom and left), 44 (all), 48, 51, 60, 61, 62, 63, 64, 66, 68, 69, 70, 71, 72, 73, 74, 76, 77, 78, 80, 81, 83, 84, 85, 86, 87, 88, 90, 91, 92, 94, 95, 96, 101, and 106

Library of Congress Control Number: 2020934208

ISBN: 978-1-938113-65-9

Item 1146

Contents

1 Introduction

PART 1

7 Exploring Math with Open-Ended Materials and Strategies

10 Chapter 1 Unexpected Treasure: How Real and Recycled Objects Inspire Thinking and Learning

14 Chapter 2 Ten Fun Activities to Engage Children in Subitizing

17 Chapter 3 Using Children's Names to Build Number Sense

20 Chapter 4 Playing with Shapes Using Pentominoes

23 Chapter 5 Creating Coding Stories and Games

PART 2

29 Embedding Math in All Areas of the Classroom

31 Chapter 6 Making Math Meaningful for Young Children

35 Chapter 7 Integrated Math in Learning Centers

46 Chapter 8 Connecting Math Concepts with Literacy Experiences *Coauthored by Erin Keith*

51 Chapter 9 Math Learning—and a Touch of Science—in the Outdoor World

54 Chapter 10 Mathematical Masterpieces: Exploring Math Through Art

60 Chapter 11 Easy Ways to Integrate Math in the Gym and Beyond

PART 3

65 Scaffolding Mathematical Inquiry

68 Chapter 12 Classroom Bird Feeding: Giving Flight to Children's Imaginations

74 Chapter 13 "Paper Shoes Aren't for Dancing!" Children's Explorations of Music and Movement Through Inquiry *Coauthored by Jennifer Bombardier*

82 Chapter 14 Mitten Measurement: Inspiring Rich Math Inquiry Through Children's Books

88 Chapter 15 Counting Kindness: How a Food Drive Inspired Rich Mathematical Thinking

97 Appendix: Reproducible Resources for Families

102 References

103 Resources

104 Acknowledgments

105 About the Author

Introduction

"Look!" I hear Pablo exclaim from across the playground. He and a few of the other 4- and 5-year-old children in my class are huddled in a group, inspecting the ground beneath their feet. As I quietly draw closer, he asks the others, "How many do you think there are?"

"Like a million!" answers Neve.

"No, a million is too many," Alessio replies confidently. "That's a big number. There aren't a million."

Aimi says, "I think maybe one hundred."

When they go back to studying the ground intently, I let them know I've joined them. "What are you looking at?"

"We're guessing how many ants!" says Pablo. He points at a crack on the sidewalk where dozens of ants are crawling.

"Whoa, that's a lot of ants! So how many do you think there are?"

As the children excitedly start repeating their guesses, I listen carefully and then ask them questions to uncover the reasoning behind their math thinking. "I wonder how we can estimate how many ants we see."

Pablo points out, "There are six around my foot!"

"And I just counted nine," adds Aimi.

"How do you know that?"

"I started counting the ones near my toes and then around my foot," Pablo responds.

"I tried to count them all at once, but they're too squirmy," says Alessio, "They keep moving and I lose track."

"It seems like it's easier to count the ants around us. What if each of us counts the ants around our feet? What could we do with that information?" I ask.

"Add the numbers together!" Aimi laughs. "Then we can figure out how many there are altogether!"

Every day, I am amazed by children's interest, confidence, and fluency in exploring math concepts in their day-to-day conversations and interactions. As a teacher of children ages 3 to 6, it is my role to engage the children in meaningful, complex math experiences that are grounded in their observations and questions about the world around them. I create provocations for them to explore, ask questions that scaffold their understanding and challenge them to think more deeply, and invite them to take risks—and make mistakes. Math is everywhere, and so are the opportunities to teach and learn it.

This was not always my approach to teaching math to young children. When I first started my career as an early childhood educator, I considered math to be just a single, isolated part of a program's curriculum. Using textbooks and worksheets, I would diligently teach the children counting and number problems during the designated time in our schedule. When that time ended, I'd move on to other subjects, leaving math behind until the next day.

Looking back, I can see that my teaching practices were rooted in my own relationship with math. In school, I was successful in early math experiences where I could rely on memorized formulas and "tricks" (*A greater-than symbol [>] is like an alligator—it eats the bigger number*). But as I got older and the math became more complicated, I struggled to understand and apply the abstract concepts I was being taught in practical situations. As a result, I lost both interest in math and confidence in my math abilities, and these feelings followed me throughout my education and well into adulthood. When I became a teacher, I tried not to let my own frustrations with learning math influence the way I taught the children, but sometimes it did.

It wasn't until years later that my mindset about what math is underwent a transformation. I read some intriguing works (see Boaler 2016; Parrish 2010, 2011) that helped me see that there were ways to teach and learn math outside of the limitations of drills and flashcards. In fact, by taking the time to intentionally look for math in the world around me, I found that the potential for engaging, meaningful math teaching and learning was *everywhere*. The songs we sing, the games we play, the everyday challenges we problem solve—it's all math! It was an epiphany, one that made me take a step back and wonder: What if, instead of treating math like an independent subject in the classroom, I tapped into the math already present in day-to-day experiences and gave the children (and myself!) a chance to play with math and see its creative applications?

Newly inspired, I started to see connections to how the inquiry-based, playful, emergent environment of the early childhood classroom could inform my math teaching approach. These are some of the most important realizations that helped to shape my practices and strategies moving forward.

Children are naturally curious about math. As children play and explore their world, their observations, questions, and ideas often incorporate math concepts, from noticing shapes and patterns in the environment to wondering how much taller they've grown since last year to counting the number of ants on the sidewalk, as the children in the opening vignette did. These are authentic and intuitive experiences with math. When a teacher steps in to connect those observations, questions, and ideas to math concepts in a concrete way, children are highly motivated to engage in creative problem solving, gather information, and build theories to support and grow their understanding.

Teachers must cultivate safe, supportive learning spaces. So that children see themselves as authentic mathematicians, they must be empowered to explore and try new things, persevere when challenged, and use mistakes as learning opportunities. A critical part of that empowerment is a nurturing environment, one where children feel that they have a voice and that their ideas are considered and valued. Children who see themselves (including their linguistic, cultural, family, and community backgrounds) reflected in the space around them and know that their thinking is being listened to develop the security and confidence to take risks in their learning, which often propels them forward in their mathematical explorations.

Collaborative learning experiences are more effective. Working together with peers introduces children to a wider range of math learning strategies than if they work alone. Children who explore mathematical situations with others see and hear different interpretations of a problem, experience alternative approaches to problem solving, and learn new representations of math ideas. When children have opportunities to talk about and play with numbers on a regular basis, they become more experienced and confident in exploring the relationships numbers have with one another (Flynn 2017; Parrish 2010, 2011).

Children are valid contributors to math teaching and learning. In the early childhood classroom, the teacher's role is not to be a math expert, but rather a learner alongside the children. Every child has her own set of ideas, interests, strengths, and experiences (or *funds of knowledge*). This means that every classroom has an impressive collective body of knowledge and skills that's just waiting to be drawn on! Honoring the classroom as a democratic space and trusting the children as creative and capable learners who are able to explore complex math concepts makes for math learning opportunities that are more exciting, energizing, and effective.

About the Book

The more I discovered the power of exploring and analyzing life through a mathematical lens, the more eager I became to share this approach. This book highlights some of the practices and strategies I have used in my own classroom over the years. To make this resource as practical as possible for other educators working with preschool and kindergarten children, it is organized into three main parts.

In Part 1, you will discover how to create math learning experiences for children using open-ended materials and strategies. Each chapter features examples that demonstrate how effective intentionally selected materials and activities can be for engaging children in math explorations that capitalize on their interests and help meet content standards.

Part 2 explores what math looks like in an emergent, integrated learning environment. When math concepts are connected to other content areas like science, literacy, and art, children are better able to see and learn math in ways that are relevant and meaningful to them. The chapters in this section highlight how to embed math concepts, materials, and tools in all areas of the classroom.

Finally, Part 3 describes how the questions children have about the world can turn into authentic, complex inquiries that incorporate or focus entirely on math concepts. As you read these examples about young children asking questions and finding answers through hands-on problem solving, it is my hope that you will be inspired to foster and generate math-rich inquiries sparked by children's wonderings.

Your role in supporting children's growth as creative-thinking, engaged mathematicians is critical. If math conjures feelings of anxiety and frustration for you, I hope the examples and ideas in this book inspire you to reimagine the possibilities for teaching and learning math in the context of playful, child-centered, inquiry-based approaches. Along the way, I encourage you to take the same advice you extend to children as they explore and understand the world through a mathematical lens:

Take risks. Make mistakes. *Have fun.*

PART 1

Exploring Math with Open-Ended Materials and Strategies

Val and Emerett, both 5 years old, search through a basket full of shells, studying and picking out their favorites as they describe what they notice.

"Oh, this one is pretty!" says Val. "Look at how the colors look like a pattern."

Emerett holds up a scallop. "I like the dark purple on this one. Purple is my favorite color. I wonder why the purple ones aren't all this dark."

After sorting through more shells, Val points out, "This one has long ridges in it. It feels bumpy."

"This one is smooth on one side," replies Emerett, "I like the way it feels soft, but it's hard."

"Hey," Val says excitedly, "we should start a collection!"

"How many shells do you need to have a collection?" asks Emerett.

"I think we have enough, but let's see. One, two, three . . ." Val counts. "Okay, we have 20 shells."

"That's a lot! We can use our collection to start a store."

"Yeah," Val agrees, "but I don't want to sell them. Let's use them for money. The purple ones can be worth one dollar and the white ones worth two dollars." She pushes the shells into two separate piles.

Emerett wrinkles his nose. "The purple ones are prettier. They should be worth more. Let's make *them* two dollars and the white ones one dollar. I'm going to get a tray to put them in."

Open-ended materials—or materials that can be explored, played with, and used in many different ways—have the potential to draw children deeply into an experience, enhancing their engagement, development, and learning (Daly & Beloglovsky 2015). These materials are transformed into whatever vision a child has in a particular moment of play—feathers are added as spices to a delicious stew, a scarf is suddenly an invisibility cloak, and tree cookies become stepping stones in a pond. Take the time to thoughtfully consider the materials in your classroom and ask yourself questions like the following:

- What potential does this object have for sparking children's interest and sustaining their learning?
- Are there real objects? Found, recycled, and natural materials? Unconventional materials?
- Does the material appeal to the children's senses?
- Will children be motivated to use the material innovatively in their explorations?
- Are there materials that are specific to the community, city, and/or geography where your program is located?
- Do the materials reflect the diverse backgrounds and cultures of the children in your class?

In addition to supporting play, open-ended materials can serve as supports in children's exploratory and organized math learning experiences. Sometimes the mathematical potential for certain materials is not immediately obvious, but when children are provided with opportunities to explore them in depth and incorporate them in their play—pretending, combining, and thinking—objects often naturally become *mathematized* (that is, framed and explored using math terms and concepts). Just as Val and Emerett pretended the shells were money for a store in the opening vignette, open-ended materials can inspire, provoke, and extend children's play, shaping the setting and story line for themes of exploration. For example, children might spontaneously notice and wonder about the many colors, shapes, and sizes of buttons in a collection and decide to sort them accordingly. Other times, math learning experiences with open-ended materials are planned and purposeful, like when children used wooden dowels to measure the length of the room.

Open-ended materials are a map that children can use to navigate the complex, abstract world of math through tangible, hands-on experiences. They create opportunities for children to test their own mathematical theories through authentic exploration in a real-world context, which in turn helps to build their confidence, interests, and abilities.

Using Math Tools

> **A few 4-year-olds are grouping toys in baskets, and I ask if they know the total amount of toys. Tamin begins to quietly mouth numbers as he counts, using his fingers as an aid. However, as soon as Tamin sees me notice what he's doing, he quickly hides his hands behind his back and looks away.**
>
> **I want the children to recognize that it's okay to use tools to help us in math explorations, but I also don't want to call Tamin out specifically. He is shy and already seems embarrassed about being "caught" using his fingers. Instead, I announce to the group, "I think I might know how many there are. If I want to add three to five, I can use my fingers to help." I hold both my hands**

> **up in front of me and model how I can use my fingers to count. "Five," I say, waving my left hand. Then, wiggling each of the three fingers raised on my right hand, I continue, "Six, seven, eight." Sending a smile toward Tamin, I explain, "Fingers are a great math tool because we always have them with us."**

Using imaginative, innovative materials in the classroom helped me rethink what a math tool might be. While they can take many forms, standard and nonstandard, they can be grouped into a few broad categories. Flynn (2017) identifies five categories, including the following two, which are the primary focus of this book:

- **Manipulatives,** or physical objects—including open-ended materials—that help children tangibly explore math ideas (blocks, geoboards, connecting cubes, even your own fingers!)
- **Representational tools,** or resources that help children organize, explore, and represent numbers (five and ten frames, number lines, hundreds charts)

These math tools invite children to become immersed in math experiences that engage their senses, helping to connect concrete actions to abstract math concepts. Take the time to introduce and model using standard math tools before having children use them. Provide children with the time and space to practice using a tool in teacher-guided small group work like a planned activity. Once children are familiar with the tool, make it available for children to access and use independently. You can continue to support and scaffold children's use of these tools while engaging in playful situations together, helping children to notice and name math learning opportunities as they arise naturally. Eventually, children will incorporate the math tool in meaningful ways in the context of their future play and explorations.

Some Terms and Tools to Know

Using and understanding math terminology can empower teachers and children. Throughout this book, specific math vocabulary and learning tools are mentioned. Here are a few brief definitions for those terms and tools.

code: a set of instructions for a computer

coding: the process of creating step-by-step instructions a computer understands and needs in order for its programs to work

five frame: a graphic organizer made up of five equal-sized boxes (each large enough to hold a counter) arranged in a row

hundreds grid: a graphic organizer made up of 100 equal-sized boxes usually arranged in rows of ten

number line: a straight line with numbers placed at equal intervals along its length

precoding activities: developmentally appropriate games and activities in which children explore some of the concepts involved in coding in ways that are meaningful to them

ten frame: a graphic organizer made up of 10 equal-sized boxes (each large enough to hold a counter) arranged in two rows of five

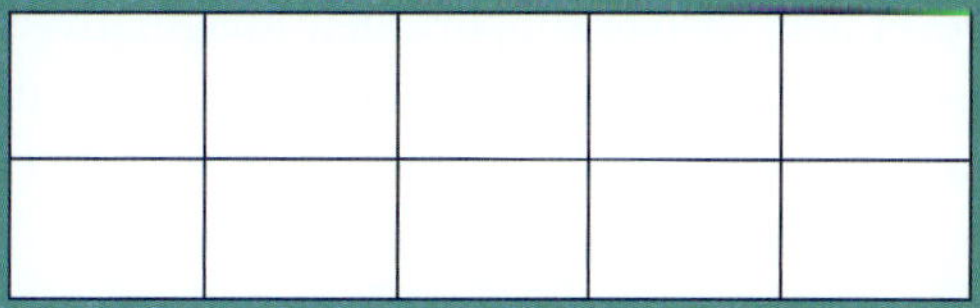

CHAPTER 1

Unexpected Treasure

How Real and Recycled Objects Inspire Thinking and Learning

> To a young child, the world is full of materials to touch, discover, and explore. To find, collect, sort, and use materials is to embark on a special kind of adventure. For adults, gathering materials means rediscovering the richness and beauty in natural, unexpected, and recyclable objects that are all around us, but not often noticed.
>
> —Cathy Weisman Topal and Lella Gandini, *Beautiful Stuff! Learning with Found Materials*

One of my favorite things to do is grab a coffee and spend an hour wandering the aisles of a thrift store, appreciating the beauty that lies within. There are so many interesting things to discover, and it's fun to imagine the stories the objects might tell if they could.

During one thrift store visit, my daughter and I came across a bag of delightful decorative spoons. As we examined them, we realized that someone had carefully curated the spoons from many different places, including the coasts of Canada, the southern United States, and even Europe! It was clear that it was once an important collection for someone. My daughter was enamored with the spoons, and she clutched them tightly to her chest. "Momma, these are so beautiful. It's so sad that these were so special to someone, and now they've ended up here!" She was bursting with questions and ideas about them: Who had the spoons belonged to? How long had he collected them? Did he travel to get each one?

One Person's "Trash"...

Young children find beauty and wonder in the most ordinary of objects. They love coming up with stories about, discovering new uses for, and experimenting with interesting combinations of all kinds of materials. After my daughter spent time exploring the spoons—researching the places they came from, using them in a tea party with her friends, and keeping a few special ones for herself—she passed them on for others to enjoy.

In the program where I teach, found objects, recycled items, and natural loose parts are incorporated into all learning areas, and children are free to tinker and play with these materials however they wish to support their imaginative stories. There are very few commercial toys in our classroom. Instead, to reflect the heart, spirit, and diversity of both the children and the wider community, many of the materials are collected or purchased from interesting places like donations

from families, yard sales, and secondhand stores. Because of this, I realized there was no one better to appreciate the rich intricacies of the spoons (the carefully carved pictures, the dangling charms, the looping handles) than my inquisitive group of students.

A few days later, I brought the spoon collection in for the children. I offered them the opportunity to explore the spoons on a table and then use them around the room. At first, most children were excited to hold each one, often noticing things that I hadn't: the gentle way a handle twisted, the funny flamingo on the Florida spoon, and how one even appeared to have a gem encased within it. As I carefully listened to and observed the children, I noticed that these treasures offered so much rich literacy potential. I heard the children read aloud the locations engraved on some spoons and use complex vocabulary to describe what they saw. I witnessed many of them become inspired to sketch the spoons in their journals or ask for books about the plants and birds featured on them so they could research them further. The children engaged in rich conversations with each other and suggested spoons for others in the class ("Elina would like the Kentucky one. It has a little horsey on the end and she loves horses!").

Over the next few days, the children incorporated the spoons into other centers, and I observed math learning explorations emerging in their work. Some children sorted the spoons based on characteristics like those with or without animals. Others lined up the spoons by length or created simple patterns with them. As they used the spoons at the sensory tables, I overheard children counting as they measured sand into bowls or reciting recipes being stirred in the big pot. It was amazing to see just how diverse and rich the experiences that incorporated the spoons were.

Soon, children began sharing their own treasures from home. One child brought in pictures of his uncle's stamp collection. This sparked an interest in how the mail system in our local community works, during which we also explored the numbers printed on stamps and how they correlated to postage costs. Another child spoke about her grandfather's coins, inspiring the children to research currency from around the world. Families also took an interest and shared even more treasures, supporting the children's explorations together at school.

Bringing Treasures into the Classroom

As you integrate found and recycled materials into your own learning space, here are some tips to keep in mind.

Introduce Open-Ended Materials

Unlike many commercial toys that have limited options for play or promote gender stereotypes, open-ended materials (like playdough, blocks, seashells, buttons, spools, and scarves, just to name a few!) are multifaceted and can be incorporated in a variety of learning centers. The opportunities for play with open-ended materials are as boundless as children's energy and imaginations. Children are active seekers and producers of knowledge, and found and recycled materials offer endless opportunities for them to invent different uses for the same materials over and over, extending their interest and evolving the nature of play over time. Whether new or familiar, children can infuse the open-ended materials with their own ideas, thoughts, and feelings as they become props in their learning and play—a crocheted doily can be a dress for a doll one morning, a flower in a bouquet the next afternoon, and a lily pad for hopping frogs to rest upon the week after that!

Provide Real Objects

Real objects (teacups in the dramatic play area, cooking utensils in the mud kitchen, safety goggles and wrenches in the science area), whether incorporated in intentionally planned provocations or spontaneous exploration and play, invite wonder and discovery. Learners who are encouraged to use objects that they see adults use at home and in the community not only feel empowered, they recognize that their learning is valued and celebrated. As a result, their confidence as capable learners is bolstered while exploring, problem solving, investigating, testing out their theories, and co-constructing knowledge with their peers.

Encourage Reusing and Upcycling

Being environmentally conscious and embracing sustainable living is more important than ever. By reusing found materials or *upcycling*—that is, creatively reusing or repurposing materials to make something new—you can inspire authentic, robust explorations that support the learning and development of the whole child *and* plant the seed for environmentally conscious behaviors and thinking in children and their families. Inspire children to find new and innovative uses for materials in order to stretch their use in the classroom. Families can also be encouraged to engage in similar actions at home and donate materials they no longer need for classroom use.

Reflect Diverse Cultures and Traditions

As an educator, you recognize and value that each child brings her own unique culture, background, and experiences to the classroom. How can you tap into these funds of knowledge to enrich children's collaborative explorations and learning? How can you have an inclusive classroom that celebrates and reflects the individuality of each child? Found, natural, and recycled objects can inspire children's creativity as they explore social, family, and cultural practices and traditions together in play situations. For example, in a community where food preparation is important, perhaps cooking utensils and recipe books might be added to the dramatic play center for children to incorporate in their pretend play. Through these interactions, children are supported as they explore and share their own culture and customs and have opportunities to be introduced to customs, languages, and values that might be unfamiliar to them. So that families are represented, you might encourage them to contribute materials for children to play with and explore in the classroom. This will help children see themselves represented in the space and encourage their acceptance of new ideas and ways of looking at the world. Children's self-confidence will grow as they feel included and recognized for the contributions they can make to the play environment, creating a safe and supportive space for all.

CHAPTER 2

Ten Fun Activities to Engage Children in Subitizing

The ability to recognize a quantity of objects instantly (like the dots on the face of a die or a domino) is known as *subitizing*. It's an important foundational skill for children's development of number sense. Playing games that involve subitizing helps children practice recognizing, adding, and subtracting quantities, which fosters future math success. Here are 10 activities that encourage subitizing.

1 Creating Equations with Dice

What You Need: Dice with numbers, operation dice with addition and subtraction signs, pencils, paper

How to Play: A child starts by rolling two number dice (dice with numbers 1 through 6) and one operation die (a die with a + or – symbol on it). Seeing what was rolled (5, 2, +), the child identifies first the numbers and then the operation sign ("A five, a two, and a plus sign").

Encourage the child to verbally create—and answer—an equation sentence ("Five plus two equals seven" or "Add five and two, and you get seven"). The equation sentence can then be written on the paper by the child or the teacher. Children who want a greater challenge can roll multiple number and operation sign dice to create more complex equations!

2 Math Subitizing Easel

What You Need: An easel, stickers or stamps, blank ten frames, a number line

How to Play: Children can visit the easel during playtime, independently placing stickers or stamps in the ten frames and writing the corresponding numerals next to them. They can also create groups of 10 using addition statements (the number of spaces with a sticker plus the number of blank spaces always equals 10).

3 Literature Connections

What You Need: Fiction or nonfiction children's books centered on counting and numbers (*Ten Black Dots,* by Donald Crews; *Pete the Cat and His Four Groovy Buttons,* by Eric Litwin, illustrated by James Dean), blank laminated ten frames, dry erase markers, wooden numbers, small manipulatives (circle counters or buttons)

How to Play: Read the story aloud to the entire class. Then arrange the book and materials on a table, and invite the children to explore counting. Children can match objects to numbers by placing manipulatives on the ten frames, then use the dry erase marker to write the numbers next to them. They can create addition or subtraction sentences on the ten frames with manipulatives or match objects to wooden numbers using the ten frames.

4 Ten Frame Names

What You Need: Each child's name written in a ten frame (one letter per square), a large ten frame, number cards from 1 to 10

How to Play: Ask children to count the letters in their names using the ten frame name cards. They can then place their name cards on the corresponding square of the large ten frame. When all children have taken a turn, the number of letters can be compared to see whose name has the most letters, the least, the same, and so on. (For more fun finding the math in children's names, see Chapter 3.)

5 Mini Catapults

What you need: Mini catapults (made from clothespins glued to wooden blocks), foam counters, sorting bowls, clipboards, paper, pencils

How to play: Children sit at a round table or in a circle on the carpet with the sorting bowls in the middle. Using the catapults, they launch foam counters, trying to get them into a bowl to score points. Record points on the paper using a tally system or ten frames.

6 Just How Many?

What You Need: A collection of small objects (shells, pebbles, gems) displayed in a clear jar, counting tools (ten frames or a hundreds chart)

How to Play: Invite children to first estimate the number of objects in the collection, then empty the jar and use one of the provided tools to count how many there actually are.

7 Race to 100

What You Need: A hundreds chart, dice, game pieces (players)

How to Play: Children take turns rolling one die or two dice and then moving their players along the hundreds chart until everyone reaches 100.

Families + Math = Fun!

Children and families can enjoy math together at home. See page 99 in the Appendix for a resource about subitizing that you can photocopy and send home.

8 Magic Ten

What You Need: Ten dice and a ten frame for each player

How to Play: Children start by rolling all 10 dice at the same time. The most common number a player rolls becomes her "magic number," and those dice are all placed on the player's ten frame. Children roll their remaining dice. Each time their magic number is rolled, that die is placed on the ten frame. Players keep rolling the remaining dice until their ten frames are filled with their magic number. Children can roll and race in pairs or groups to see who can fill their frame with the magic number first.

9 Rolling and Recording

What You Need: A die, paper with the numbers 1 through 6 written horizontally in a row, a pencil

How to Play: Children roll the die and make tally marks next to the corresponding number on a piece of paper. Or, instead of making tally marks, they can practice writing the number when it is rolled. The children can be challenged to see which number is rolled most frequently in a particular number of rolls.

10 Domino Parking Lot

What You Need: Mats created by the teacher with the numbers 1 to 12 written in evenly divided spaces in a grid, dominoes

How to Play: Encourage children to sort dominoes by counting the dots and placing them in the corresponding space in the grid. For example, the domino that has a total of six dots would be "parked" in the 6 space. Children can also park two dominoes whose total number of dots equals the number in the space (a 6 domino and a 5 domino could be parked in the 11 spot).

CHAPTER 3

Using Children's Names to Build Number Sense

> Remember that a person's name is to that person the sweetest and most important sound in any language.
>
> —Dale Carnegie, *How to Win Friends and Influence People*

One of the first and most important words that a child will learn to read and write is her name. Children learn to recognize their names and those of their peers in a literacy-rich environment where names can be incorporated into daily routines and activities. In addition to empowering children's literacy skills and cultivating a sense of community and collaboration in the classroom, exploring and manipulating names is packed with rich math potential! Here are some easy ways to help children connect early numeracy concepts using names as a foundation.

1 Name Wall

What You Need: Each child's name written in a ten frame (one letter per square), 10 blank ten frames, stickers or markers, tape or adhesive putty, paint chips/swatches (optional)

How to Play: Lead the children in creating a wall display that organizes their names by how many letters they contain. First, use stickers or drawn symbols to mark the 10 blank ten frames to represent the numbers 1 through 10. Affix these ten frames to the wall to act as categories for the name wall. Once each child has a ten frame with his name written on it, have him count how many letters are in his name and determine which ten frame number his name should be placed under. (For a fun pop of color, use recycled paint chips as the ten frame for their names.) Once the name wall is complete, children can study and use it as a reference to see who in the class has the same number of letters in their names and who has more or less letters.

2 Graphing Our Names

What You Need: Number cards for 1 through 10, connecting cubes or other linking manipulatives

How to Play: Invite children to explore representing the number of letters in their names concretely with linking manipulatives like connecting cubes or LEGO bricks. For each letter, children use one cube (the name *Kala* would be represented with four cubes). Arrange the number cards in a clear space on the carpet or a table. Once each child assembles a line of counting cubes to represent her name, have her place it under the corresponding number card. This will create a point of data on the graph. Once all the children have added to the graph, analyze it together and talk about what you see.

3. Clothesline Names

What You Need: Yarn or twine, name cards, photos of children, alphabet cards, clothespins, pencils, paper

How to Play: Create a clothesline with the yarn, and display children's photos and name cards, leaving space between each. Place the clothespins and alphabet cards in baskets along the clothesline. Encourage children to find their photo and name card (or the photo and name card of a friend) and to use the clothespins to clip the alphabet cards in the correct sequence to spell that name. Have each child count the number of alphabet cards they used and record that number on paper.

3 Clothesline Names

What You Need: Chart paper, markers, name cards

How to Play: As part of your morning message, choose one child to be the mystery star. On chart paper, write a clue about the child ("He has an orange kitten named Tangerine") and an outline of that child's name using blank boxes for each letter. Be sure that each box reflects the relative size of each corresponding letter. Prompt children to explore all the elements of the mystery name as they try to guess who the star is: "How many letters are in the name? What do the letters look like? Are they little or big? Tall or short? Is there a recognizable pattern to the letters? Is it the same number of letters as your own name? Less or more letters?" When the children guess who they think the mystery star is, hold up that child's name card to see if it matches the blank boxes. Once they guess correctly, write the mystery star's name in the blank boxes.

5 Decorating Names

What You Need: Each child's name printed in large letters on paper or spelled out with wooden letters, a collection of small manipulatives (buttons or pom-poms), stickers

How to Play: Children use small objects or stickers to cover the printed or wooden letter spelling out their names. Once finished, have them count how many objects they used. Record their answers and explore whose names used the most or least materials. You can also encourage children to make patterns as they decorate their names in different ways!

6 Scavenger Hunt

What You Need: Handouts that feature children's photos with a ten frame beside each one, clipboards, pencils

How to Play: Children search the classroom (name wall, name cards, cubbies, documentation displays) for their classmates' names. When the child finds a name, she writes it in the ten frame beside the corresponding photo on her handout and notes how many letters are in that name.

7 Five Frame Names

What You Need: Each child's name written in a five frame or a ten frame (one letter per square), small manipulatives (bottle caps or washers)

How to Play: Encourage children to explore their five or ten frame and then put the small manipulatives on the frame, one per letter. Have each child consider the number of letters in her name related to the anchor of five or ten, depending on whether her name is five letters or shorter and she is using a five frame or six letters or longer and she is using a ten frame. As everyone explores their names as a group, engage the children in more complex number talk to help build their fluidity and confidence working with numbers: "Whose name is equal to five? Whose is less than five? Whose is more than five? How many letters less than five is Aki's name?"

8 Name Addition

What You Need: Paper, markers, small manipulatives (large beads or acorns)

How to Play: Draw outlines of each child's name using blank boxes. As in the Guess Who? activity, have each box reflect the relative size of the corresponding letter (a lowercase *m* would have a box that's short and wide). Have the children write in their names in the outlines you provide and then ask them to consider the letters in their names and sort them in two different groups (tall and short, wide and skinny). Encourage them to use manipulatives to represent the number of letters in each of these two groups and count the total number of manipulatives. Model how these numbers can be written in the form of an addition sentence. For example, Sam has one tall letter and two short letters in his name, so his addition sentence is "One plus two equals three."

CHAPTER 4

Playing with Shapes Using Pentominoes

Pentominoes—or polygons made up of five squares that are equal in size and connected side to side—are versatile math manipulatives that encourage children to explore composing and decomposing shapes. They

- Promote geometric awareness and spatial reasoning skills (including logic, symmetry, reflection, rotation, and design)
- Foster thinking about measurement (including length, width, area, and perimeter)
- Integrate number sense
- Encourage cooperation and collaboration among children
- Help children develop focus and perseverance

When introducing pentominoes to children, use math language to help them identify the characteristics of each individual piece (there are 12 standard pentomino shapes in one set) and the set as a whole. Incorporate children's observations and ask guiding questions to highlight the number of squares, sides, and corners that each piece has. Model how the pentominoes fit together in different ways, with or without spaces between them. Prompt children to study and talk about the differences they see when comparing one pentomino shape to another, and help them see that no two are the same.

Here are six activities that incorporate pentominoes that help children build a richer understanding of complex math concepts. For each, be sure to carefully observe children's actions and conversations so you can highlight and scaffold the math concepts and language they are exploring.

1 Open-Ended Exploration

What You Need: Multiple sets of pentominoes in a basket

How to Play: In a large, clear working area, have the basket of pentominoes easily accessible for children to freely explore the pieces and place them together in different configurations during playtime. Children may fit them together like a puzzle, order them in patterns, or create picture representations (butterfly, face, flower). As children become more confident, encourage them to work together to create bigger, more complex patterns and designs.

2 Fill the Frame

What You Need: Multiple sets of pentominoes in a basket, trays or picture frames

How to Play: Children can work individually or together to fill the area of a tray or picture frame with pentominoes without leaving any empty spaces. Having various sizes of trays or picture frames differentiates the activity and makes the experience easier or more challenging for children.

3 Follow the Leader

What You Need: Multiple sets of pentominoes in a basket, masking or duct tape

How to Play: Place tape in the middle of a large, clear working area to create a horizontal or vertical line of symmetry. Explain to the children that one side is yours (the leader's) and the other is theirs. Place a pentomino on your side of the line and then invite children to take turns finding and placing an identical piece on the opposite side, flipping it so that it is in a symmetrical position. Continue adding pentominoes to your side, one piece at a time, and having children do the same until an intricate symmetrical design emerges. When children have a firm understanding of the game, have them take turns in the role of leader.

4 Last Piece!

What You Need: Two sets of pentominoes (one set for each player), a large tray or picture frame

How to Play: Each player takes a turn placing pentominoes in the picture frame, once piece at a time, until they are both left with only pentominoes that do not fit. The player with the fewest remaining pentominoes in his pile wins the game. Encourage children to experiment with different sizes of frames—the smaller the frame, the more challenging the game becomes.

5 Grid Puzzles

What You Need: Multiple sets of pentominoes in a basket, grid paper, markers

How to Play: Invite children to fit pentominoes together, without leaving any empty spaces between them, to create four-sided shapes. Have them trace each pentomino in their shape with a marker on grid paper. Children then trade the grid puzzle they've created with another child and try to complete their partner's puzzle. Encourage them to compare the grids and note any similarities or differences, like the number of pieces needed to fill each grid or how many squares total there are on a grid. Older children can be encouraged to create easy addition or multiplication facts that represent how many total squares make up their grid puzzle.

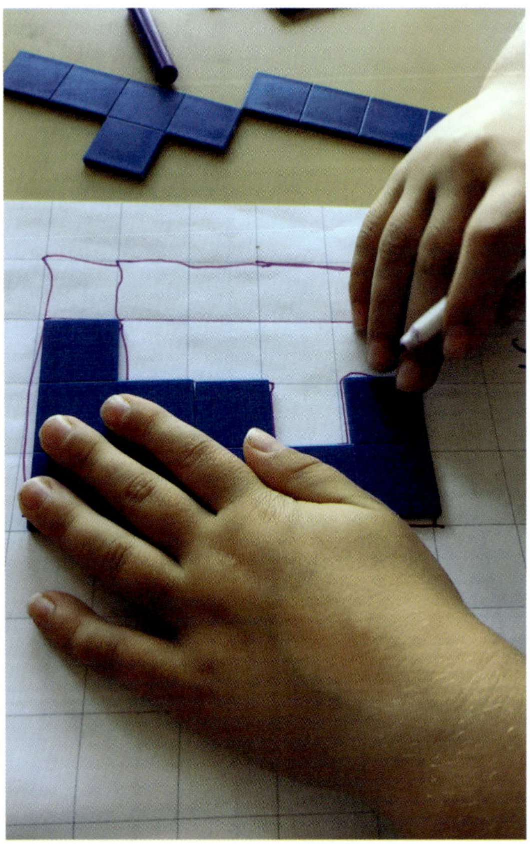

6 Code a Story Path

What You Need: Multiple sets of pentominoes in a basket, paper, writing materials, small props to represent characters from a chosen or created story

How to Play: Decide if you and the children will retell a favorite story or create your own, and gather the small character props (if retelling *Hansel and Gretel,* you would need character props for Hansel, Gretel, and the witch). Determine the start and end points of the path (these correspond with the beginning and end of the story), then have the children create a complex path that connects these points by arranging pentominoes end to end. For an initial run through, place the character prop at the start point and move it along the path until reaching the end. Repeat the path, but this time, ask the children to code the path—that is, to describe and record step-by-step instructions (the direction of each move, how many squares to move) the character needs to make during each move to successfully reach the end point. For an extra challenge, add or remove pentominoes in subsequent rounds or create a more complex path with multiple dead ends. (For more on coding games, see Chapter 5.)

CHAPTER 5

Creating Coding Stories and Games

Coding stories and coding games are playful, hands-on ways for children to explore and experiment with early coding. They offer opportunities for interactions and collaborative learning. If coding is new to you, you will find that it builds on many early math and literacy concepts you are familiar with.

What Is Coding?

Coding (or programming) is a basic language of the digital age. It involves the process of creating step-by-step instructions a computer understands and needs in order for its programs to work. Gaming systems, tablets, cars, cell phones—even washing machines!—all use coding to function properly. Early coding, or precoding, offers children experiences that integrate communication, thinking, and problem solving. These are 21st century skills that are valuable for children's future success in our digital world.

There are many learning activities preschool children already engage in that relate to coding. I started to think about some aspects of precoding that would be meaningful to preschoolers, and I began to develop a series of activities where children create their own coding stories. Children learn about creating commands, or using directional language, when they make and use maps to locate a treasure during dramatic play and when they give each other instructions while playing with toy cars. Activities related to creating and following commands can be game-like in nature, such as moving a game piece a specific number of spaces on a grid (think Candyland or Chutes and Ladders). Children are naturally drawn to adventurous experiences in which they can explore movement, direction, and location. Many of the same concepts supported in these activities, such as spatial reasoning and number sense, are a foundational part of coding that children can practice without a computer.

Early coding is fun and exciting. The earlier children explore the basics of coding, the more easily they will be able to learn, understand, and apply coding later in life. Many preschoolers are just the right age to begin learning!

What Do Children Need to Know?

Coding is more accessible to young children than you may think. To participate successfully in precoding activities, children should be introduced to directional language like *up* and *down, backward* and *forward,* and *left* and *right*. They need to become familiar with how to count using ordinals (*first, second, third*) and understand one-to-one correspondence. Some children will have mastered these concepts and others will still be learning. If children are just beginning to understand and follow directional language and use ordinals, the precoding activities that follow will reinforce that learning.

Activities involving location and movement, such as playing board games and giving directions to peers in pretend play, offer some background knowledge and provide entry points for children into coding activities.

Why Introduce Coding to Preschoolers and Kindergartners?

Coding is like a game. It's very engaging for children who enjoy telling stories and using grids and maps. Children ages 3 to 5 are able to create drawings of maps that represent relationships between objects and places. When we incorporate programming into early learning settings, we immerse children in versatile activities that align with standards in multiple areas, such as math, problem solving, communication, and literacy (Umaschi Bers 2018). Precoding activities offer children opportunities for interaction and collaborative learning as well.

Coding Stories and Games Use Math

Coding involves spatial reasoning and logic. Helping children think about direction, location, and movement strengthens their math skills. Children need to count the number of spaces they must move and indicate in which direction they are headed as they relate objects and locations to one another. Coding integrates a number of math skills into one meaningful experience.

Coding Stories and Games Encourage Problem Solving

When activities like coding provide a challenge for the user or when there are several ways of solving the same problem, children can consider multiple paths for arriving at a solution and choose the most efficient strategy. In these activities, planning ahead is the key to success.

Coding Stories and Games Strengthen Communication

When a child codes, she needs to indicate directions to another person (or to a computer). Her communications need to be succinct—that is, they cannot be wordy. When two children play coding games, the child receiving the instructions needs to be an active listener in order to move the coding pieces in the correct way. Clear communication and turn taking are essential to coding.

Coding Stories and Games Promote Literacy

Coding games often tell a story—there is a path the programmer takes to move an object through different obstacles and settings, eventually arriving at a destination. Children's imaginative stories or favorite retellings can be used as the base for coding games. This helps children develop vocabulary and a greater comprehension of texts. Children can make connections between literature and their lives. They can even change the outcome of a favorite story by exploring different endings in the coding activity. Teachers can lead innovative read-alouds with a gaming board by changing the sequence of a story or reinventing the ending.

Books that Introduce Coding

Here are a few books that offer ideas for interactions and collaborative learning around coding in your classroom. You can use these titles yourself to better understand computational thinking and to introduce the foundational skills of computer coding to children.

Mapping Penny's World
By Loreen Leedy (2000)

Lisa's teacher invites the class to make maps of anywhere in the world. Lisa chooses her bedroom. When Lisa includes her dog Penny's bed, it inspires her to map Penny's world! This book includes directional language, legends and symbols, and colorful illustrations of detailed overhead maps that lead the reader to different places in Lisa's home and yard. It is an excellent resource to introduce foundational concepts and vocabulary that will prepare young children for later coding opportunities.

Try this: Many children ages 3 to 5 are able to create drawings of maps that represent relationships between objects and places. Customize mapmaking and precoding activities for the class based on your knowledge of the children. Depending on the range of children's developmental stages, you might use only a few sections of this book.

Coding is meaningful to young children when it is prompted by a familiar story. Use this book as an introductory text to help children think about a series of events from start to finish and how a map can tell the story. Or children can sequence the events in Lisa and Penny's world and discuss how the characters get from one place to another. Introductory coding lessons can help children program these movements. The book can be a guide for those children who are ready to experiment with coding.

Hello Ruby: Adventures in Coding
By Linda Liukas (2015)

Ruby has a wonderfully rich imagination, and she and her animal friends go on many adventures together. As Ruby explores her world, she introduces coding concepts such as problem solving, step-by-step planning, and patterning. Later in the book, children can use the coding fundamentals included throughout the story in fun activities. There are two sections in the book. The first is suitable for younger children just beginning to learn about coding, as they consider Ruby's adventures and the different choices she makes. The second part may be more appropriate for children who are ready to experiment with some initial coding ideas and activities.

This book is an informative resource for teachers who are new to coding. The introduction addresses families, but teachers will learn a lot about the kind of thinking children need to practice for later use in the digital world.

Try this: Use the book to get ideas for meeting children where they are developmentally. The chapters are very short and are great conversation starters. Encourage children to use what they know about Ruby and her world to solve the challenges in the book or to create their own adventures! Be inspired to try some of the activities in a small group.

If I Were a Wizard
By Paul Hamilton
Illustrated by Simon Howe (2017)

When Ralph's teacher asks the class what they hope to be when they grow up, Ralph's ideas are different from his classmates'—Ralph dreams of being a wizard who can change the world. Coding concepts are interwoven throughout the story to help make them easy for young children to understand and access. The book connects coding concepts to children's immediate world. It is an appropriate selection for children who are familiar and experienced with coding terms and strategies.

Try this: As children become comfortable with coding terms and actions, they can begin to apply them to concrete coding activities using commands and a grid. You can refer back to the book and practice the coding activities in a real-life context like coming to school in the morning.

(Adapted from Pecaski McLennan 2017)

Getting Started

First, make a grid. You can create a grid of squares using masking tape on the floor or on a tabletop. A grid can also be drawn on a large plastic tablecloth, making it easy to use, transport, and store. For children who want to explore coding further, make a large grid by tracing black lines on grid paper. This way, a child can draw his own coding story and take it home to share with family members.

Choose your story and gather the corresponding props. Decide if the children are going to retell a favorite story (*The Three Little Pigs, Goldilocks*) or invent their own tale. For example, if the children choose to explore *The Gingerbread Man,* they will need a mini gingerbread figure, a fox, and pictures to represent the places the gingerbread man travels to in the story.

Ensure that children understand the story line and the path they are coding. Determine the starting and ending points of the path the character is going to take. In *The Gingerbread Man,* the character starts in a cottage, then moves down a lane, through a field, and into the woods. He finally arrives at a stream, where he meets a fox. If the children are retelling a favorite story, be sure their retelling is in the proper sequence and makes sense. If they are inventing their own story, children can place different settings or obstacles around the grid for a character to maneuver through.

How to Play

1. Set up the grid. The teacher or children place the character at the starting point on the grid and put the settings or obstacles on the grid in the order they occur.

2. Establish the commands. As a group, the children determine the commands for moving the character along the grid. Directional arrows (which can point up, down, left, or right when placed on the grid) are visual symbols to represent the commands. Children can easily make and use lots of arrows on mini cards or sticky notes.

3. Choose a programmer and a gamer. The programmer—or programmers—gives the coding commands. The gamer is the child—or children—who follows the coding commands and moves the figure on the grid.

4. Code the story! The programmer is now responsible for giving commands—"Go up 1," "Go forward 2," "Go down 3,"—to the gamer so that the gamer can move the character along the grid from start to finish. (The first few times a group codes, an adult might model the roles for the children.) As a new command is followed, the coder places a symbol card in each square the gaming piece travels through, representing the direction of the move on the grid. Code all the moves until the ending point is reached.

Encourage and support children so they can complete the entire path, giving and following correct directions without skipping squares on the grid. Ensure that the directional arrows are left in place during the game to represent the path the character has taken. When the character arrives at the final destination, the path is complete! Other children can change the story and have a turn.

Extending the Coding Play

Once children become confident with early coding, they can use it independently in many different ways. Coding naturally and easily complements other learning opportunities in preschool and kindergarten classrooms. Children can

- Code outdoors by drawing grids on the pavement with chalk and retelling stories using corresponding props
- Create their own coding stories using grid paper by establishing a story line, determining a starting and ending point, and drawing coding symbols directly on the grid
- Create scavenger hunts for one another, placing treasures in the squares of a grid; children can direct one another using a coded message to reach a certain square and then claim the treasure
- Incorporate elements of precoding into their dramatic play, both indoors and out, by drawing or placing grids on top of maps and directing each other using commands to move to different locations (going on an adventure, searching for treasure)
- Invite family or community members to visit the classroom and share ways adults use coding in real life

PART 2

Embedding Math in All Areas of the Classroom

My kindergarten class is cutting shapes from colorful construction paper and gluing them onto cardboard. As they work, I notice that most of the children are arranging shapes to create picture representations like houses, cars, and people, using circles as wheels and heads and rectangles for things like chimneys and bodies. A few add more intricate details to help depict what the shapes are supposed to be.

Six-year-old Adia, however, is doing something entirely different. As I draw closer, I can see that she has painstakingly cut out a number of squares, all blue and equal in size, and is gluing them in neat rows and columns to fill the entire surface of the cardboard. Curious about her work, I ask her to describe what she's doing. "I want to see how many squares it takes to fill this," she says as she points to her work. "Then I'm going to flip it over, use another color, and see if I need the same number of squares for that side."

I'm amazed! Adia's interest in exploring the area of each side of the cardboard to determine if they are equal is an abstract concept that we have not yet explored in our classroom. I am eager to continue observing her and see what she discovers. I ask Adia if she will share her artwork during our next circle time meeting and mentally begin to plan how to extend this interest in measurement in future art activities.

For math to make sense to children, they need to be able to see, touch, hear—even smell and taste!—it in contexts that they are familiar with or interested in. How does this happen? As an educator, it is important to be receptive and responsive to the diverse skills, interests, and ideas that children have. When you acknowledge and embrace the many different things children already know, like, and can do, you can use them as stepping stones to teaching children new information and skills. Looking for those connections and finding ways to honor and support children through integrated, multifaceted explorations help you to enrich everyday activities with authentic math.

While content areas (math, science, literacy) are often thought about separately, no area of learning can or does happen in complete isolation. In fact, weaving together concepts across the curriculum often leads to more meaningful and enjoyable learning experiences. Math is a universal language that lends itself to being integrated in all areas of the classroom. At first, you might see the math in some experiences and activities more easily than others, but with a little practice at mathematizing, you'll soon find rich math potential is waiting to be discovered all around you: In the cozy reading area, where children discuss the patterns they notice in their favorite books. Among children's constructions on the large carpet, where their attention to symmetry and balance are observable. In the art area, as children experiment with shape, color, and texture. During the complex games children invent and as they play with equipment in the gym and outdoors. The possibilities are endless!

Provide children with plenty of hands-on opportunities to explore and share their thinking about math in a variety of ways (measuring the size of a painting, constructing a tower with a limited number of blocks, determining who ran faster during the relay race). When children understand and have fun with the math concepts they learn, they are more inclined to naturally engage in math learning opportunities, which builds their confidence and agency as mathematicians. By encouraging children to share their observations, ask questions about the world around them, and play with math without fear of failure and in ways that are meaningful to them, you empower them to engage in increasingly complex and dynamic math experiences in the future.

The chapters in this section explore some ideas and strategies for how you can create a classroom environment that inspires math-rich thinking and learning. Chapters 6 and 7 offer suggestions for how you can encourage math explorations through the questions you ask and conversations you have with children as well as the materials you select and integrate in learning centers. Chapters 8 through 11 go into greater depth about how you might embed math concepts across different content areas. As you read, consider each area of your early childhood program and think about how math is already present in it and how math learning can be highlighted throughout it.

CHAPTER 6

Making Math Meaningful for Young Children

Children are natural mathematicians. They push and pull toys, stack blocks, and fill and empty cups of water in the bathtub. All of these activities allow young children to experience math concepts as they experiment with spatial awareness, measurement, and problem solving (ETFO 2010; NAEYC & NCTM [2002] 2010). Young children learn as they describe, explain, and consider the ideas from their immediate environment: "Am I as tall as Yancey? How can I find out? I know! We can both stand next to each other in front of the mirror."

Early math is not about the rote learning of discrete facts like how much 5 + 7 equals. Rather, it's about children actively making sense of the world around them. Unlike drills or worksheets with one correct answer, open-ended, playful exploration encourages children to solve problems in real situations. Because the situations are meaningful, children can gain a deeper understanding of number, quantity, size, patterning, and data management (Grossman, n.d.). For example, it is easier to understand what *six* means when applied to a real-life task such as finding six beads to string on a necklace or placing one cracker on each of six plates.

Creating a Math-Rich Classroom

Research suggests that preschool classrooms can be the ideal environment for learning about math (ETFO 2010). Children sort materials into corresponding bins at cleanup time, explore patterns and shapes while creating at the art table, tell time while using the visual schedule to predict which activities come next, and measure when they squeeze their bodies through the climber on the playground (ETFO 2010).

Preschool classrooms also celebrate curiosity and risk-taking as children engage in inquiry-based exploration at various learning centers and outdoors. Interesting items in the environment encourage children to find answers to their questions and solve problems across all curricular domains. Children measure as they clap out the beats to music. They repeat rhythmic patterns as they dance. They describe, sort, and count objects in the science center and look for patterns while on a nature walk. They count rungs while climbing up the ladder to the loft. Many familiar children's songs, stories, and poems contain mathematical messages that help familiarize children with counting, measuring, and patterning. For example, children can count along with "One, Two, Buckle My Shoe" and "Ten Little Monkeys Jumping on the Bed."

In addition to offering blocks, buttons, and other loose materials to touch and explore, teachers can ask open-ended questions that promote problem solving as well as probe and challenge children's mathematical thinking and reasoning skills (Ontario Ministry of Education 2016). Such questions are not meant to elicit correct answers but rather to engage children in open-ended conversations that promote high-level thinking, such as "What do you notice about these objects? How could we sort them?" One of the foundations of play-based learning is that the teacher is active in the play, asking questions and adding knowledge and insight. The teacher learns together with children throughout the inquiry process.

Every preschool classroom needs to be rich with materials that encourage math exploration and learning. A well-stocked math and manipulatives center includes found objects (shells, stones, bread tags, sticks) as well as purchased materials. It can include photos of completed geoboard creations or of children sorting coins in the dramatic play center. There might be narratives of children's learning, such as transcripts of children's comments and conversations, and artwork featuring pattern or shape exploration. Teachers can post documentation of math learning as a way of encouraging children to reflect on past experiences and motivate them to plan and revise future

Supporting Dual Language Learners

Children who are dual language learners (DLLs) can learn math concepts and skills without being fluent in their second language. Much of the meaning is found in the right materials. If families send familiar items from home to the classroom, the children will know the name and function of the items in their home language. They can use this prior knowledge as a foundation to help them learn math. For example, young children may not understand how to sort plastic shapes, but they already know it is important to sort the baby's socks and daddy's socks in separate piles—a math activity that has real-life meaning in any language.

ones. These visuals can inspire even deeper, more connected learning. They help children maintain their focus on a particular topic, refine and expand their ideas, communicate their learning to others, and reflect on their experiences before making new plans.

Encourage Children to Play Mathematically

Young children need to see themselves as capable mathematicians. Child-guided and child-focused explorations and teacher-guided math activities help children practice and consolidate their learning. This helps them feel confident about what they know and can do. Although many preschoolers learn some math concepts on their own, it's important for teachers to include math in authentic experiences, resulting in a deeper understanding by children (ETFO 2010).

In addition to creating a rich math and manipulatives learning center, teachers can encourage children to use math tools and strategies in all areas of the classroom. Children might use a set of plastic links to measure their buildings in the block center, use play money to pay for a train ticket in the dramatic play center, and use rulers to measure the growth of spring bulbs in the science center. Take a set of scales outdoors so children can figure out who found the heaviest rock. Using math tools for real-life tasks frees both teachers and children to act spontaneously, resulting in richer interactions and a calmer learning environment (Wien 2004).

Families + Math = Fun!

Children and families can enjoy math together at home. See page 101 in the Appendix for a resource about exploring and playing with math concepts that you can photocopy and send home.

Children need the freedom to use materials in authentic ways as well as the freedom of time and space to deeply engage in math. The daily routine should include plenty of time for uninterrupted play so children have the time they need to work on sustained tasks of interest. This allows children to explore materials thoroughly, often resulting in more complex and evolved experiences over time. If a child spends all of his time at one learning center, he is not missing out on learning opportunities elsewhere. Instead, his deep connection to the center is often indicative of rich learning. Teachers can model

the use of other materials at the center, such as using writing materials to draw plans for a structure to be built, or pose challenges that encourage the child to think beyond her play, such as "How tall can you build this tower before it falls?"

To support learning, it is important to encourage children to communicate their explorations and findings. Teachers can establish a routine through which children share their experiences at group time. For example, a child might explain how he built a structure with blocks, do a dance with repeating steps, or share a photo of a complex pattern made with colorful buttons. While circulating through the room, a teacher might notice high-quality work and suggest that a child share it with her peers during group time. The child making the presentation grows in confidence and her peers may want to try the experience themselves.

Conclusion

Most children enter preschool knowing a lot about math. In a safe and supportive classroom, they will feel comfortable taking risks and engaging in self-directed problem solving. Weaving math into all areas of the curriculum will heighten children's play experiences and allow all learners to experience success. Children will soon see themselves as capable mathematicians who apply their skills in a number of ways. Their growing math skills, confidence, and interests will serve them well in school and life.

CHAPTER 7

Integrated Math in Learning Centers

Math teaching and learning experiences can happen anywhere in the classroom, not just in the defined math center. By strategically embedding rich mathematical materials, tools, and resources in *all* learning centers, you create countless opportunities for children to engage in authentic, playful math exploration and problem solving.

To foster a learning space that encourages and supports play-based math learning, here are some things to consider:

- Give children the freedom to explore self-directed areas of interest. Then scaffold children's learning during those explorations with open-ended questions, comments, and suggestions that prompt them to consider their interests and observations through a mathematical lens.
- Seeing how math can be used in real contexts with concrete materials is essential for children to build an understanding of math concepts. A math-rich space encourages children to notice the world around them and use resources innovatively to play, create, and problem solve.
- Provide a wide range of familiar and new materials, including open-ended materials (blocks, playdough), natural materials (bird nests, pinecones), found and recycled materials (shoeboxes, milk jugs), and real tools and objects (scissors, screwdrivers). Some materials should be available throughout the year, but others should be swapped in and out to keep children's interest and to inspire new ideas.
- All children are capable, creative mathematicians. When challenges arise, emphasize a child-directed approach to problem solving, either independently or with peers.
- The classroom is a democratic space where all children's voices are important and valued. Honor the knowledge, skills, and experiences children bring with them into the classroom as well as their questions; each child's unique individual knowledge enhances the class's collective understanding of math.

This chapter explores ideas for materials you can integrate in some of the core learning centers found in many preschool and kindergarten classrooms and how they can nurture math-rich play and learning.

Yoga

The Art Center

Art making provides children with endless opportunities to express themselves, make decisions, use their experiences as inspiration, communicate in different ways, collaborate creatively, and deepen their conceptual understanding. It is a personal and emotional process that helps children connect their thoughts with actions. Art can also help children who are pre-emergent readers and writers express their ideas using symbolic languages like painting, drawing, and sculpting. As children experiment with a variety of materials and create artwork, your intentional guidance can help them discover the math in their art. For example, art making can complement ongoing inquiries happening in the classroom. If children are exploring flowers, they might study and recreate the patterns they discover in the petals in a painting; while studying insects, they could create a collage illustrating their observations about the number of legs insects have compared to the number of legs arachnids have.

Examples of Basic Materials	Potential Math Learning Opportunities
• An assortment of paper (construction paper, newspaper, wrapping paper) • Sketchbooks and clipboards • Markers, crayons, and pencils • Paint and paintbrushes • Clay or playdough and various cutters and rollers • Scissors, tape, and glue • Droppers and food coloring • Found and recycled materials (bread tags, bottle caps, feathers, cardboard boxes and tubes, pipe cleaners, beads)	• Design bracelets and necklaces with intricate patterns using string and beads • Create figures and scenery using geometric shapes • Measure the material needed for a particular artwork (length of yarn needed to make a necklace, what size frame is needed to display a drawing • Create new colors by counting, adding, and mixing together food coloring (1 drop of red and 1 drop of blue makes 2 drops of purple) • Sort and organize various materials in the center (place all paint brushes in a tall vase, separate scissors and glue in separate bins, group similar color crayons together)

The Building Center

The building center invites children to design and create structures in a hands-on way. As part of this process, children brainstorm, predict, plan, collaborate, test, and so much more. The building and creating process is rich with opportunities for exploring math concepts, including counting, ordering, spatial reasoning, shapes, measurement, symmetry, and patterns. This area often becomes the heart of a classroom, and it's not unusual to see children bring materials from the different learning centers into this space in order to heighten and layer their play—after all, Antoine wants a baby doll from the dramatic play center to enjoy the crib he just made with blocks!

Examples of Basic Materials	Potential Math Learning Opportunities
• Blocks in a variety of sizes, shapes, colors, and materials (wood, foam) • Other building toys (LEGO bricks, KEVA planks, K'NEX) • Standard and nonstandard measuring tools (measuring tapes, protractors, yarn) • Maps and blueprints • Natural materials (sticks, rocks, shells) • Found and recycled materials (ceramic tiles, picture frames, corks, cardboard) • Marbles or small balls • Roleplaying props (tool belts, hard hats, construction safety vests) • Dramatic figures and vehicles (people, animals, trains, trucks) • Materials for writing and drawing (clipboards, grid paper, pencils, markers) • Fiction and nonfiction children's books on architecture and engineering	• Sort materials by size, shape, color, or texture • Explore symmetry and patterns in the structures children design and build • Count and measure different materials • Compare the height and width of structures (seeing which tower is taller) • Measure which car rolled the farthest or fastest down a ramp • Explore how the shapes of different building materials relate to one another (spatial relationships) and can be flipped or rotated (spatial transformations) while building a structure • Write ideas and draw pictures that represent architectural plans • Problem solve when creating complex mazes and marble runs • Exercise logic and puzzling skills while fitting loose parts together

The Discovery Center

Children are naturally curious, and the discovery center taps into that curiosity by providing children with opportunities and resources to observe and explore the world around them. In addition to using their senses and existing knowledge and experiences, materials and tools encourage children to engage in investigations where they ask questions, test their ideas, make observations, and communicate their findings to others. Math is built in to these emergent, inquiry-based explorations, and the materials you select to be included in this learning center can highlight math concepts both through intentional learning experiences (a long-term project tracking whether plants left in the sun or the shade grow taller) or invite more open-ended explorations (discovering during play what happens when a tower that is very tall and narrow is built without a wide, solid base). These discoveries help children construct personal and collective understandings of the world in which they live.

Examples of Basic Materials	Potential Math Learning Opportunities
• Living things (plants, an aquarium) • Natural materials (pine needles, tree cookies, sea glass) • Standard and nonstandard measuring tools (balance scales, connecting cubes) • Observation tools (magnifying glasses, binoculars, thermometers, stethoscopes) • Magnets and magnetic items (paper clips) • Mirrors • Tweezers • Sorting trays • Birdfeeder • Safety goggles • Overhead projector and accompanying materials (transparencies, clear shape pieces) • Materials for writing and drawing (pencils, paper, markers, crayons) • Fiction and nonfiction children's books related to a specific area of study or interest	• Match colors to natural objects (a red paint chip to a red flower) • Sort and classify materials using common characteristics (shells by size, feathers by color) • Identify and document patterns in nature (spiderwebs, honeycombs, the life cycle of a tree, markings on animals) • Copy and create patterns using found and natural materials • Measure and record how tall a plant grows over time • Predict and record how long it takes for a seed to germinate • Count and record the number of seeds found in a fruit or vegetable • Record and graph the temperature outside • Count and record the number of birds that visit a feeder during a given period of time

WE
RECYCLE

The Dramatic Play Center

Children love to imitate scenes that they see in their daily life at home and in the community. The dramatic play center offers them both a dedicated space and real objects for them to reenact those experiences and imagine others. Traditionally, this center is organized as a house, but it can transform depending on the needs and interests of the children (vet's office, restaurant). Dramatic play uses roleplaying and improvisation (for young children, this is spontaneous, unscripted play and make-believe) to help them explore, analyze, and navigate personal feelings and social issues. This kind of play engages children's emotional, cognitive, and verbal skills and can serve as a framework for hands-on mathematical problem solving in meaningful, real-life contexts (determining how many cups of flour are needed to bake a birthday cake, drawing a map for a road trip). Provide objects that are culturally meaningful to the children in your setting and accurately reflect their everyday lives.

Examples of Basic Materials	Potential Math Learning Opportunities
• Dressup clothing and accessories (hats, scarves, shoes, stethoscopes, aprons, sunglasses, feather boas, purses and wallets) • Furniture and pretend appliances (table and chairs, shelves, stove, telephone, camera) • Dolls, puppets, and related accessories (bottles, blankets, strollers) • Kitchen play props (plastic food, dishes, utensils, cookbooks) • Shop play props (pretend money, cash register, receipts, reusable shopping bags) • Materials for writing and drawing (pencils, paper, markers, crayons)	• Sort materials when tidying up (all the babies sleep in the crib, all the cups go on the top shelf) • Read or write a recipe and then follow it while pretend cooking • Write a grocery list with the number of each item needed • Set the price for items in a store or restaurant and count pretend money to purchase them • Write down or dial a phone number • Retell the sequence of events in a play situation using ordinal words (*first, then, next, finally*) • Set the table using one-to-one correspondence • Create a schedule or record appointments on a calendar

The Literacy Center

The literacy center encourages children to engage in reading and writing activities that empower them as communicators. With chairs or soft pillows and plush toys, the reading corner provides a quiet, comfortable space for children to read alone or with one or two other children. Books can be sorted and organized by common characteristics (color, subject, author) to help children find their favorites easier. At the writing table, children have access to materials that invite them to explore and take risks in the symbolic representations of language that they are most comfortable using, from scribbling and making letter-like shapes to writing words with conventional spellings. They are able to explore letters and the sounds they make using different manipulatives, draw blueprints for a structure they are building, and write letters and mail them to one another using a class mailbox system.

Examples of Basic Materials	Potential Math Learning Opportunities
• A variety of diverse fiction and nonfiction books and magazines • Paper and envelopes in an assortment of sizes and colors • Materials for writing and drawing (pencils, markers, crayons) • Journals and clipboards • Alphabet cards • Scissors, tape, and glue • Magnetic numbers and letters • Stamps and ink pads • Name cards • Book posters	• Count along or find shapes and patterns in storybooks • Measure and fold paper to fit inside of an envelope • Conduct surveys and make graphic organizers to record and interpret the results (using a T-chart to determine how many people want to come to a birthday party before creating invitations) • Write numbers when addressing letters to one another • Count the number of letters in a word or the number of words in a sentence • Compare the number of letters in two different words to determine which is "bigger" • Write and/or illustrate books and number the pages

CHAPTER 8

Connecting Math Concepts with Literacy Experiences

Coauthored by Erin Keith

Literacy provides a powerful foundation for inquiry and learning. Children research topics of interest by exploring books and other printed materials; make marks, write words or sentences, or draw pictures to represent their evolving understanding of the world around them; and use robust vocabulary to describe their ideas to one another. Because communication—that is, reading, writing, drawing, and speaking—is a critical tool across all areas of the curriculum, the literacy center often becomes the central hub of a classroom, especially when children are exploring highly engaging topics. In our program, for example, when the children were raising painted lady caterpillars, the literacy center was busy each day as the children read about butterfly life cycles, recorded time-lapse videos, drew their observations of the metamorphosis process, and communicated using new terminology they were learning in the context of the experience (*pupa, chrysalis*).

Math and literacy naturally complement one another, and math concepts become all the more meaningful and relevant when they are incorporated into classroom inquiries. Consider the caterpillar study. As the children learned about growth and changes that the caterpillar experienced over its life cycle, they were also learning about a kind of pattern. When children explained what happened during the caterpillar's transformation into a butterfly, they wrote numbers to represent certain timestamps in the videos they recorded or more broadly described what they observed using ordinal words ("First, the caterpillar hatched from the egg. Next, the caterpillar ate leaves and grew a lot longer and wider . . .").

Here are some suggestions for how you might recognize and intentionally embed math learning opportunities in the literacy experiences children have in your classroom.

Create an Environment that Highlights Math Concepts

Post child-created and adult-modeled print throughout the entire classroom (numerals, number words, and math terms posted near the math center; co-created math anchor charts in the literacy center). In the literacy center, include a variety of tools and resources that inspire children to integrate math ideas into their explorations. All kinds of materials and activities can invite inherently mathematical thinking. Compasses and protractors inspire composition and decomposition of shapes. Graphic organizers like Venn diagrams allow children to compare and contrast the attributes of animals or objects, while T-charts help them record and organize survey results. Before mailing a letter to a friend, children must fold the paper so that it fits inside of the corresponding envelope, encouraging exploration of measurement, shape, and spatial relationships.

As with all areas of the classroom, organize materials in the literacy center by common characteristics. Children can begin to explore and develop a more complex understanding of how objects can be grouped together. For example, they might discover that while it is more useful to group index cards into different baskets by size, markers and crayons are more easily found when sorted by color. Even books are sorted, perhaps by shape and size as children figure out how many of each type of book (tall, short, wide, narrow, thick, thin) can fit on the different shelves while displaying them in a way that is appealing.

It is important to use and encourage math talk, which fosters children's development of a positive attitude and confidence about math. As children talk about their day or describe a piece of work they've created, help them think about the math that's part of what they're seeing and doing, and use language that they can learn and in turn use ("I see that you are writing a letter to your friend. I wonder how you'll determine what size envelope you need to use," "How will you decide how many papers you need to staple together in order to have enough for the book you are writing?"). Work together with children and listen carefully for opportunities where you can chime in to scaffold any math talk that emerges in their play and inspire deeper thinking about math concepts.

Explore Math Concepts Through Children's Books

Rich fiction and nonfiction children's books often inspire some of the most interesting math inquiries in the classroom. Books can be integrated in so many different ways and different places in the classroom. They can be read individually or in small or whole groups as read-alouds. Books

should be present throughout the classroom, not just in the literacy center, where they can enhance an ongoing interest the children have or inspire a completely new one!

While some books deliberately have a math focus, almost any high-quality book can spark math learning. A nonfiction book about insects, for example, might result in children exploring insects with symmetrical markings (like butterflies and moths) which can lead to an investigation of copying patterns and creating new ones. Books about interesting structures from around the world (such as *Dreaming Up: A Celebration of Building,* by Christy Hale) can be used as blueprints while children build with big blocks and explore geometrical concepts.

Repetitive (and often silly!) texts are always a favorite with young children. Their predictable nature makes the books an interactive experience as children follow along and recite the words with you during a group read-aloud. Through intentional questions and comments, you can help children recognize that a pattern is what makes the text predictable and that they can recite the story along with you because they have learned that pattern. Books such as the There Was an Old Lady series, by Lucille Colandro, illustrated by Jared Lee, or *We're Going on a Bear Hunt,* by Michael Rosen, illustrated by Helen Oxenbury, can also be retold through roleplaying or with props. The physical, hands-on play and material manipulation when dramatizing a sequence of events from a book grounds the children's understanding of how patterns work. Once children are comfortable recognizing and copying patterns in storybooks, encourage them to extend the pattern from the story or even to create their own pattern book by using the materials in the literacy center to write or draw their ideas. (For a more in-depth look at how children's books can inspire rich math inquiries, see Chapter 14.)

Sing Ten Frames

Singing is an engaging way for young children to explore many concepts of literacy, including rhyme, vocabulary, and comprehension. With a little intentional thought, songs can also enhance math learning. There are countless opportunities to incorporate familiar songs and fingerplays into the daily routine, including whole group time and transition times. Sing-alongs like "Five Little Monkeys" and "Ten in the Bed" have math built in, helping children practice counting forward and backwards in stable order to and from five and ten. Songs that use anchors of five or ten are particularly engaging for children because they can dramatize the characters using their fingers. To further mathematize these experiences, you might use a visual five or ten frame for the children to follow along with while singing. Easily transportable versions can be made using a metal baking tray, tape, and magnets. As children sing along and use their fingers to count, you can represent the corresponding number on the ten frame, helping children to see the number and eventually learn to

subitize it. Once children are comfortable with this activity, they can take turns leading the song by manipulating the ten frame. These materials can also be available during free play so that children can continue to explore the math with manipulatives in different contexts, such as while playing school in the dramatic play or literacy centers. (To see a video of singing ten frames in action, visit www.mrsmclennan.blogspot.com/2015/11/singing-ten-frames.html.)

Play with Words

Pre-emergent readers and writers benefit from participation in early literacy activities that encourage exploration of concepts of print like letters and words work together to create a message that we can understand and directionality (that English-language books are read from left to right and top to bottom). While introducing concepts of print to young children, there are many strategies you can use to simultaneously explore math concepts like sorting, number sense, and measurement. For example, the syllables in children's names can be clapped out and counted. Together, you might explore the size and shape of letters or how words can be sorted by the number of letters they contain. Sentences can be measured by counting and comparing the words in each ("Which word is the longest?"). Classic games can be given a twist that's both literacy and math related. You and the children might play a version of I Spy, searching the room or the pages of a book for words composed with a certain number of letters ("I spy with my little eye a word that has three letters!").

Fostering Math Learning with Dual Language Learners

To provide all children with equitable learning opportunities, early childhood educators must recognize, honor, and build upon children's individual experiences and knowledge including their languages, cultures, and families. With such diverse backgrounds laying the foundation for development and learning, it's no surprise that each child you teach comes to the classroom with her own unique approach to exploring and making sense of math concepts. While play and inquiry can help to make math learning experiences more inclusive, additional supports and strategies are key to supporting dual language learners.

Math, like many content areas, depends on language. Number words, for example, are different in each language, and in some languages, a different numeral system is used. Consider researching or engaging families to learn some basic math vocabulary in the children's home languages. Not only will learning how to count to 10 in a child's home language better help you support her, it can also provide an opportunity for other children to learn how to count in a second language. Math-related print materials posted in the classroom should also include number words and numerals in children's home languages. Engage in rich math conversations with dual language learners using math terms in both English and their home languages during activities and routines throughout the day (discussing the shape, number, and size of different foods during snack time).

Embed books that relate to math concepts in the various home languages of the children you teach throughout the classroom (a Spanish-language book about shapes in the construction area, a Korean-language counting book on the loose parts shelf). If the book is not also available in English, you might add English translations with sticky notes. This allows all children to see themselves in their learning environment and encourages dual language learners and English-speaking children to explore books together.

Conclusion

Learning to read and write is an exciting milestone for young children! Cultivating a print-rich space where literacy scaffolds math learning and helping children explore the concepts of print through a math lens will empower them to think more deeply as readers, writers, *and* mathematicians.

About the Coauthor

Erin Keith, EdD, OCT (Ontario Certified Teacher), is an assistant professor at Bishop's University in Lennoxville, Quebec, Canada. She teaches pre-service teachers how to holistically support children's self-regulation and mental health through intentional, play-based practices while using the environment as a powerful third teacher. Her research and professional practice explore early childhood wellness and trauma-informed, wrap-around supports in school-based settings.

CHAPTER 9

Math Learning—and a Touch of Science—in the Outdoor World

Young children strive to understand the world around them. Outdoors, they can spend hours engaged in rich, imaginative play using nothing more than the natural materials they find. How many times have you emptied treasures from the pockets of a preschooler or kindergartner? Pebbles, pinecones, feathers, and leaves are wondrous and hold so much potential for math and science inquiry!

Discovery Outdoors

The great outdoors is filled with rich opportunities for math learning—with a connection to science—that can interest and engage children in real-life problem solving. Children need to spend large blocks of time playing outside; classes in our program spend at least one hour outdoors every day. It's amazing to see how children's mathematical explorations and scientific observations outdoors inspire our indoor learning. We once spent several weeks examining the characteristics of the plants and trees around our school! The children's initial interest in trees sparked hours of reading, researching, drawing, sculpting, and dramatizing different forest adventures. Outdoor exploration also complements emergent math *and* science.

Here are some suggestions for integrating learning opportunities into play in natural spaces and for connecting outdoor explorations to children's learning. Get ready to look at the outdoor world through a mathematical lens with an eye for science!

Rethink Your Surroundings

The design and contents of the outdoor play space are important. When children have access to different found and recycled objects, including logs cut from fallen trees, tires, and large stones, their imaginations guide activities. Allow children to control the direction of their play while engaging in big body movement. Complex learning occurs through children's authentic planning and problem solving. They can line up tree stumps by size to make stair steps, roll them into certain locations to define a space, use them as platforms to jump from, and arrange them with stones and tires to create an obstacle course.

Include areas that contain sand, water, pea stone gravel, and other sensory materials. Place old tables outside to give children plenty of working space, and use items like trays and pots for organization. Add tools such as tubing, buckets, shovels, measuring cups, and spoons to encourage children to use math when talking about their actions and making plans for collaborative play. Engage children in math conversations as they use the materials to explore quantity and movement. Have children dictate recipes for mud stew, discuss measuring ingredients for leaf salad, or price out and buy supplies from the "store."

Use Math Tools Outdoors

Many outdoor play areas include traditional vehicles and toys like tricycles, wagons, trucks, and balls. By adding math tools to this collection of materials, you encourage children to play in unexpected and more complex ways. Offer laminated ten frames, sorting trays, recycled checkerboards, hoops, and number lines. Teachers can scaffold activities as children incorporate the tools into their play. Loose natural materials like stones and acorns can be placed on ten frames to encourage counting; flowers can be organized on sorting trays by color and size; and sticks can be counted using a number line. Use chalk to draw a large hundreds grid on the pavement, or add natural loose parts along with dice or number spinners to a checkerboard to inspire children to create math games together.

Be Responsive to Children's Observations and Wonderings

When playing with children, pay attention to their comments about the outdoor space. Note the questions children have ("How big is this tree?," "How fast can you run?," "Why are all those birds flying together?") and encourage and support their emergent projects. For example, in our program, we bring a tablet outside so the children can document and photograph events they observe (like the leaves changing colors and the geese migrating south for the winter) and then share their observations during whole group conversations.

Teachers can clarify children's questions and help them plan an investigation to discover answers. Books and websites can provide background knowledge and introduce appropriate math and science tools and vocabulary. Children see firsthand why math is important when they observe authentic purposes for it in their world. Vocabulary and traditional problem solving have more meaning when they're applied to real situations.

When the children in my classroom were curious about the large trees in our playground, we practiced using standard tools (measuring tape) and nonstandard tools (connecting links) to measure the trunks. The children learned and remembered the word *circumference*—the distance around something—because it mattered to them in that moment. I hadn't intended to do a measurement activity with the children that day, but because I was responsive to their interests, we experienced a rich mathematical exploration that incorporated many learning standards.

Explore and Sort Collections

Encouraging children to keep collections of natural materials like feathers, shells, or seed pods is an easy way to integrate math and science into outdoor play. Children can identify common characteristics (size, color, shape) and use them to create sorting rules to group items. Children can also bring materials indoors for in-depth exploration and play. We keep collections in the various centers in our classroom so that the children can incorporate them into their daily activities. The objects can be added to geometric designs at the art table, placed on the light table for patterning, or examined with a magnifying glass in the science center. We encourage families to search their yards and neighborhoods for more materials, adding variety to what has already been gathered at school.

Notice Patterns

Patterns are everywhere in nature! From the unique markings on animals to the cycle of the changing seasons, there are many ways children observe and document repetition in our world. Natural patterns can be complex, requiring children to think abstractly and interpret their surroundings in different ways. When a child observes the seasonal stages of leaf growth on a tree, there are many possibilities to connect their observations to math, science, and literacy. Unlike more common activities for patterning that might have only a few correct answers (coloring in boxes on a worksheet, beading on a string), exploring natural patterns encourages children to think deeply about what they are observing: Why does the pattern occur? Will the pattern change in any way? How does the pattern affect other parts of the natural world?

Some patterns are more difficult to identify, and because the outdoor play area offers different experiences each day (depending on the weather, human interactions, and animal visitors), there are always new things to discover, interpret, and talk about.

Children can look closely at patterns by using tools like magnifying glasses and microscopes to examine rocks, leaves, and spiderwebs. In our classroom, the children took digital photos to create books about patterns they found in nature. They also copied and extended the patterns they saw outside in their artwork using pastels and loose parts. They communicated new understanding by blogging and tweeting with others about their nature-inspired math work. The children enjoyed connecting with other classes from around the country.

Conclusion

Children find joy in the wonders of nature. Tap into this excitement to create authentic, engaging, and accessible math and science experiences outdoors. When we open our hearts and minds to the possibilities of nature, we make learning fun and relevant for all children right in their own backyards!

CHAPTER 10

Mathematical Masterpieces

Exploring Math Through Art

For young children, art is a language. Drawing, painting, and sculpting help them investigate the world and communicate their understandings. Children practice real-life problem solving when they use different materials and tools to turn their ideas into creations.

To support children's math learning, offer a range of teacher-guided and child-led math opportunities connected to art. There are so many possibilities for math in preschool and kindergarten classrooms. Here are some ideas to inspire you!

Create an Art Center with Mathematical Possibilities

Our art area is in a sunny corner of the classroom, framed by large windows. We have a variety of colorful purchased, found, and recycled materials. Their textures and colors are functional and beautiful. Here children can transform their ideas from play into reality (making a baseball cap for a doll) or follow up on observations of the natural world (painting birds they see at the playground feeder). Teachers model art techniques, talk with children about their creations, and help children when needed.

During our whole group gathering time, we often introduce new art-making techniques, materials, and tools to the children. When we add items (smaller paintbrushes, food coloring, droppers) to the art center, we demonstrate the tools' possible uses (create a watercolor painting with a dropper) and how to care for them (rinse the dropper when finished). This way, children can work independently in the center.

As with all learning centers, math plays a large part in the art space's organization. Materials are displayed in clear containers and are sorted by color, texture, and size. Arranging the materials by characteristics requires children to become sorters at cleanup time.

Consider using children's artwork to decorate the classroom and display math projects. Strands of colorful beads strung by the children in intentional patterns can hang in the windows. Jars of colored water can be visual reminders of the different shades of colors they created. When children want to display their paintings, they engage their math skills to find appropriately sized picture frames. They measure their artwork with measuring tapes or rulers and then find picture frames that match those dimensions.

Notice and Name Math

Children enjoy using materials in innovative ways. In the art center, we encourage them to freely explore and create. For example, when long, colored paper strips were added to the center, the children used them in different ways. Many made patterns by gluing the strips onto sheets of paper. Some glued them end-to-end to make longer strips that became snakes or sashes or snow skis. Others twisted the strips into spirals or folded them to create complex 3-D structures and patterns.

We teachers observed, questioned, and reflected with children about their creative processes. Through thoughtful scaffolding, we helped children see connections between math and art. Teacher–child conversations focused on mathematical aspects of the children's creations: "I see that you are folding the paper into a lot of little triangles. The length of the strip is getting shorter." The commentaries made math concepts concrete for the children. They were able to talk about geometric concepts and use math vocabulary like *repeating, sequence, height, width, length,* and *area.*

Play with Symmetry

Experimenting with art making can lead to surprising results. In our classroom, children folded and pressed their wet paintings together, delighting in the new forms their pictures took as the colors blended. This activity helped the children explore *symmetry* firsthand. They saw that when they folded their wet paintings down the middle, each half was a mirror image of the other.

Symmetry is a math concept that is often present in nature—fruit, snowflakes, and even human faces have symmetry. As the children became more skilled at recognizing this characteristic, they represented their new knowledge in their art. When the children created butterflies by painting on transparent paper, they were careful to first draw the shape of the wings and then design symmetrical patterns on them; for example, five circles on each wing.

During another activity, children looked in a mirror to sketch their faces. Children who did not draw their features symmetrically commented that their self-portraits were not accurate. They knew that their facial features needed to be balanced. With time and experience, the children began to point out symmetrical objects in the school (double front doors, the class's pet crab, teachers' faces) as well as outdoors in nature (leaves, animals' faces, pairs of birds' wings).

Identify Patterns in Nature

Natural objects offer other rich mathematical and art possibilities. We regularly use them for observation and as models for drawing in the art center. Tree cookies have concentric circles. Zebras and chameleons have striped patterns. Spiral patterns can be seen in nautilus shells, pinecones, and pineapples.

We encourage children to spend time gently handling natural objects (flowers, acorns), examining them, and looking for patterns. They use tools like magnifying glasses and tweezers to look closely. The children record their observations in sketchbooks and add details to their paintings, drawings, and sculptures. When particular objects, such as feathers, aren't available, we provide photographs of them. Discussing patterns they see with classmates and representing patterns in their artwork help children deepen their understanding of patterning.

Solve Math Equations with Color

In children's art, the process of creation is more important than the final product. One of the discoveries the children made while creating artwork was that colors can combine to create new colors. Children can mix colors by using droppers to tint water with food coloring or to dilute tempera paint with water. They can melt different combinations of colored ice cubes made from tinted water or overlap various colors of plastic film on the light table. Or they might create lighter or darker shades by adding white or black paint.

Playing with color helps children make rich mathematical connections. We encourage children to think of equations when mixing colors.

Children making green with food coloring can explain the procedure as an equation:

For orange, when melted:

Children are inventors, creating and naming unique colors:

As children invent their own colors, we add their recipes to a master list of color equations. Then others can recreate the colors by following the equations.

Once knowledge of the primary colors and recipes for secondary colors are established, children can invent variations by changing quantities:

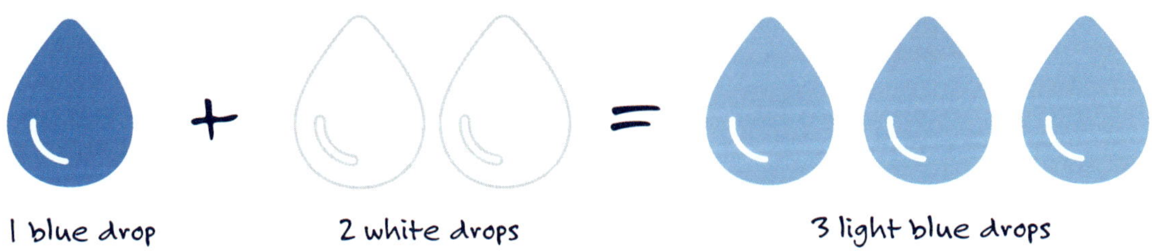

Incorporate Shapes

Like color, *shape* is one of the seven building blocks of art—in addition to line, form, value, space, and texture. Shapes can be geometric (circles, squares, triangles) or organic (leaves, vines, clouds). Children manipulate and play with shapes in the art center. They sketch and paint 2-D shapes or cut and paste them from a variety of materials (felt, foam, wrapping paper). Budding architects build with 3-D shapes using recycled boxes, cans, and cardboard tubes.

Teachers can engage child artists in geometry discussions by noticing and naming the identifiable shapes in children's art. They can point out regular and irregular shapes and talk about how the shapes enrich the children's creations.

When a child paints an outdoor scene, shapes help make the figures recognizable (circles for flowers, triangles for evergreen trees). In construction and sculpture, shapes matter. For example, a square base for a cardboard tower works, but a sphere would not. When children use patterned geometric designs (unit blocks, pattern blocks, tangrams) for art creations, they use *spatial reasoning*—that is, children consider how shapes fit together.

Conclusion

Creating art is one way children make sense of their world. When educators join children in art activities, they can integrate math learning into the creative process in meaningful ways. Doing so helps children experience the usefulness and beauty of mathematics, one masterpiece at a time.

CHAPTER 11

Easy Ways to Integrate Math in the Gym and Beyond

Math activities aren't just for inside the classroom. With a few simple tools—dice, rulers, number lines, and number spinners—children can infuse math into big body activities. Here are nine easy activities for integrating math in the gym and beyond!

1 Strengthen Children's Number Sense with Dice

Use two dice—a standard one with the pips one to six and another with the name of a different exercise on each side (jump, push-up, squat). Roll the dice and encourage children to perform the number of exercises shown (six jumps). Challenge learners by rolling another standard die and encouraging them to add the pips together for the total number of exercises to be performed.

2 Inspire Children to Record Their Data

Ask children for suggestions of exercises they like to perform (laps around the gym, jumping jacks) and set the timer for one minute. Challenge children to keep track of how many of each exercise they can perform in one minute. Then, ask them to work together to calculate the total number of exercises they completed as a group. Record this number and refer to it the next time the children participate in this activity.

3 Race Across the Gym Using Number Spinners

Children can play parts in a life-size board game when number lines are created on a floor surface. Use a ruler to partition the number line into equal segments and mark them off with tape. Two or more children can race from one side of the gym to the other. Each player starts at 0, takes a turn spinning and identifying the number, and then jumps, skips, or hops the corresponding number of places on the number line. The first player to reach the end of the number line wins.

4 Create a Measurement Challenge

Encourage children to throw an object (a beanbag, a paper airplane) as far as they can and measure how far it flies. Children can problem solve how to calculate the distance when an object is thrown beyond the length of one measuring tape or ruler.

5 Jump or Hop Along a Number Line

Number lines can easily be made on the floor using tape or durable plastic strips. Children can use them to record how far they can jump. Each player starts at 0 and jumps as far as she can. The activity can be modified by having a child roll a die and attempt to jump to the corresponding number on the line (if a 5 is rolled, the child jumps to the 5 on the number line). Challenge children by having them roll two dice and add to see where they need to land. When children are first learning to add, they can hop the equation: if a 2 and a 3 are rolled, the child first hops to the 2, then hops three more spaces and lands on 5 to represent the equation 2 + 3 = 5.

6 Encourage Children to Calculate Quantity Quickly by Playing Grouping Number Games

Have children mix in the gym or outdoor space by skipping, hopping, or tiptoeing to music. When the music stops, call out a number and have the children quickly form groups (if 3 is called, children scramble to form groups of three).

7 Take Advantage of the Gym's Large Space to Do Playful Ordering Activities

Encourage children to move creatively to music. When the music stops, call out a specific way for the children to arrange themselves. This can include a teacher-directed challenge (shortest to tallest) or can incorporate math props (children wearing numbers have to order themselves from 1 to 10 or children have to form two lines—one with even numbers and one with odd numbers).

8 Foster Spatial Reasoning Using a Large Grid that Children Can Maneuver Within and Around

Use masking tape or sidewalk chalk to create a grid large enough for children to move in. Determine a starting and an ending position. One child stands on the start position and is guided around the grid by another child who calls out directions (move up one space, move right three spaces). To make the activity more challenging, place obstacles on the grid that must be maneuvered around.

9 Have Children Build Shapes with Their Bodies

Write the names of various shapes (circle, triangle, diamond, parallelogram) on index cards. Select a card and encourage children to work together in small groups to make the shape on the card with their bodies, either standing together or laying on the ground. Vary the experience by increasing or decreasing the number of children in each group or by calling out more complex shapes.

PART 3

Scaffolding Mathematical Inquiry

"Look, Mrs. McLennan—we found tracks!"

I turn in the direction of the voice calling out to me and see Mark and Chayna, two of my 6-year-old students, deep in discussion as they gesture at the fresh snow. "Wow! What do you think made these?" I ask.

"I don't know," responds Chayna, "but it must have been running real fast. See how big the spaces between the footprints are?"

"Yeah, and I think it had four legs," Mark adds. "Look at the pattern of the different feet!"

"Good observations! Let's see what other tracks we can find."

As the children begin to search the area, I take a photo of the prints in the snow with my tablet for later exploration. This interest in tracks is a great entry point for deeper exploration of measurement and patterns, and my thoughts are already racing with ideas for how to extend it into a math-rich inquiry.

Some of the most meaningful and complex learning experiences children have are self-directed, teacher-guided inquiries that stem from an observation, wondering, or question children have about the world around them. Children arrive in their first year of school with a wealth of experiences and ideas that shape how they observe and interpret the world around them. Through engaging hands-on play and teacher-guided learning experiences in the classroom, you help children connect what they already know and can do with new ideas and skills.

Sometimes an inquiry happens spontaneously through authentic learning moments in a rich environment, like when Mark and Chayna became interested in the tracks they discovered in the snow. Other times, learning experiences are intentionally planned and implemented by the teacher (Dietze & Kashin 2018). For example, you might notice a few children in your class love navigating cars and trucks along the rug that depicts a neighborhood road map. Tapping into that interest, you might ask if the children can come up with ways to give a driver detailed directions for how to get from one place on the map to another. Some inquiries may last a day or two while others might result in a month-long exploration. Regardless of how long an inquiry lasts or the question from which it originates, these projects help children to apply, analyze, and interpret knowledge in authentic, hands-on situations.

Inquiry-based learning encourages children to solve a problem or answer a question that is important to them (Heick 2019). Rather than follow a linear, pre-determined curriculum, educators that encourage inquiry in the classroom use the observations and questions that children have to guide curriculum development. Throughout the inquiry process, children have opportunities to experiment and explore, observe what happens, gather information that supports (or disproves) their ideas, and reflect on and share their findings with others. With the guidance of an educator who holds a magnifying glass up to the things children are curious about, an inquiry can also forge a pathway for rich math thinking and learning.

As a teacher, you spend time observing, documenting, and engaging with children as they play and explore. By being attuned to children's curiosities about the world and engaging with them in authentic math explorations, you transform into a co-learner alongside the children. Honoring the children's interests and scaffolding their learning through those interests help educators navigate and embed required standards and assessment strategies into a context that's both motivating and meaningful for children (Wien 2014). When children express interest in a topic, you can nurture and mathematize their curiosity through the materials you provide, the questions you ask, and the extension activities you plan. These math experiences help to positively shape the way children view and interpret math in their lives and build their mindset.

Inquiries require children to use math in different ways that can be thought about as steps.

Step 1: Understand the Problem or Create a Question to Be Explored. Potential for inquiry can happen anywhere, anytime, so flexibility in planning and implementation is key. You must be prepared to improvise alongside the children while simultaneously planning for future learning opportunities (Wurm 2005). In this initial step of the inquiry process, children and educators ask themselves a number of questions to help refine the focus of their inquiry, including

- What is the problem?
- What do we hope to find out?
- What do we already know?
- What don't we know?
- What information might be still be missing?

Step 2: Create a Plan. Once there is a clear, concrete question or problem to explore, create a plan for how to proceed with the inquiry. Ask children to consider what materials and tools are needed to move forward with their explorations. You can help deconstruct the problem into smaller steps and provide resources to support their data collection, brainstorming, and planning.

Step 3: Enact the Plan. Children then take the information, questions, and planning they have done and put it in action. As they hypothesize, test, evaluate, retest, and refine their work, they are researchers, working toward a final outcome. With each step children can analyze and reflect upon their findings. What are they noticing? Are they closer to answering their question or solving the problem? Do changes need to be made? How can they record their observations? Do the results seem reasonable? If not, does the plan need to be modified and tried again? As they work, new questions and problems might come up and inform the children's work along the way.

Step 4: Reflect and Share. Encourage children to reflect on the entire experience and to consider whether they achieved what they set out to accomplish. Was their initial question answered? Did new questions emerge as a result of their work? As a wise mentor once advised me, the point of good research is not to gather answers but to spark more questions to explore. If satisfied, children can then share their findings with a greater audience—the rest of the school, their families, or even the wider community! Effective ways to share new learning with others might include creating a blog post together, putting together a documentation display to exhibit in a public place, or illustrating a book to be added to the school library.

CHAPTER 12

Classroom Bird Feeding

Giving Flight to Children's Imaginations

Little did I know on that frosty morning in October, when I first hammered the pole into the ground and affixed the gazebo-style bird feeder on top, that I was lighting the imaginations of the children in my classroom. Always an avid backyard birdwatcher, I hoped to sneak peeks through the classroom windows at the feathered visitors that might come during the school day. The almost 80-year-old school had well-established, diverse landscaping in the yard. Just outside the classroom's three-tiered windows stood a round formation of evergreens and tall shrubs, perfect for attracting and protecting birds at a feeder.

A Perfect Place for Bird Feeding

Despite being an architectural beauty, the school was designated by our school board as eligible for special funds and services because of its high levels of student transience, low socioeconomic levels, numerous dual language learners, and high rates of student absenteeism. At the time, I taught a large class of 27 4- and 5-year-olds. Many were new immigrants who were likely to quickly transfer to other sections of the city once their families became established. The children differed greatly in their experiences, skills, and interests, which made our program and daily interactions exciting and engaging. There was never a dull moment.

The education program was primarily a play-based, emergent approach to learning. Children engaged in self-directed, collaborative, exploratory, and arts-based activities for most of the day. Although I enjoyed integrating elements of nature into learning centers, I originally set up the bird feeder for personal purposes. I expected that some children might show an interest in it, but I had no intention of including it in the regular classroom curriculum. I had just hoped to continue a hobby to which I devoted much of my spare time.

I was right in my earlier assumption that the lush and diverse greenery would attract a variety of birds to the feeder—but much to my amazement, the children's interest in the bird feeder was instantaneous! The day I put up the feeder, our class discussion focused on what it was and how it would attract birds. I learned that many children had never experienced hanging an outdoor feeder. Most of the children lived in cooperative, subsidized housing or large apartment buildings and didn't have access to a big yard with large trees. As that first day with the feeder progressed, many hopeful faces spent their choice time standing at the windows, waiting for our first guests to arrive. Children are often eager to learn about the environment, and an emergent, inquiry-based approach is ideal for capitalizing on these interests and engaging children in environmentally friendly practices that will hopefully last a lifetime (Wien 2014).

Within a few weeks, the feeder attracted some colorful local birds, including house finches, goldfinches, sparrows, and even a few blue jays. As new birds arrived, we consulted birding websites or asked the teacher-librarian in our school for help finding reference materials to identify the birds and what they liked to eat. Researching the birds together with the children allowed them to see me as a fellow learner who wanted to continue to expand her own knowledge (Wien 2014). I was modeling lifelong learning.

Classroom visitors would think that a superhero had arrived by the level of excitement and noise that happened when the children noticed a new bird. Surprisingly, many children continued to spend choice time watching the feeder, and they often called out to their peers when they spotted a new arrival. Every child appeared interested in the activities at the feeder, and no one missed the opportunity to see a new bird.

An Enriched Literacy Program

Autumn gave way to winter and the birds continued to excite the children. What had begun as a personal activity for me was now the focal point of our classroom, providing inspiration for child-led inquiry and discussion. The colder weather brought numerous birds to the feeder, and they usually ate one large bag of seed each week. Children began to incorporate their observations into their play in other classroom centers. Many children who usually were not motivated to engage in language activities began to draw pictures and attempt emergent writing to describe the birds and the changing seasons. During visits to the school library, the children checked out nature books, and at their request I read many fiction and nonfiction books about birds at story time.

As the children noticed more varieties of visiting birds, they became intrigued with bird names and characteristics. This inspired them to write a class letter to a nature sanctuary during our shared writing time. We requested information on local wildlife. The sanctuary wrote back to us and included a large poster depicting native Ontario birds, which we promptly posted on the wall next to the window. As we spotted new birds, we recorded their names and the date on a chart. All winter the children kept a log of what they noticed, similar to winter birding counts conducted by serious birdwatchers. Along with the poster came informational brochures on local wildlife, which I placed in the language center; many children became engrossed with the detailed pictures of new animals. This nonfiction material supplemented our reading program well, as the children were incredibly motivated to conduct their own research by analyzing the brochures and then creating their own brochures about birds in the language center.

By midwinter the children still spent much of their choice time watching the feeder, aided by the new binoculars that hung next to the window. I noticed a gradual change in the conversations children were having at the windows. What was once just talk of the birds slowly evolved into a more sophisticated, personal dialogue, full of philosophical references to life and personal interests. Children engaged in rich discussions about what they loved about the birds and how the birds had inspired them to engage in other activities in the classroom. They elaborately planned for what they would play together during outside time and how they could continue to watch the birds during these activities. They dreamed of the possibilities of flying like our birds and visiting exotic lands. As children stood by the windows, they began to engage in meaningful conversations with each other. The bird watching had achieved what I had tried earlier to develop through planned circle time discussions. The communal experience of bird watching provided a natural, secure atmosphere for children's discussions to grow and flourish.

Bird-Related Resources for Children

1. *About Birds: A Guide for Children,* by Cathryn Sill, illustrated by John Sill (1991)
2. *Don't Let the Pigeon Drive the Bus!* by Mo Willems (2003)
3. *Backyard Birding for Kids: A Field Guide and Activities,* by Fran Lee (2005)
4. *Feathers: Not Just for Flying,* by Melissa Stewart, illustrated by Sarah S. Brannen (2014)
5. *Mama Built a Little Nest,* by Jennifer Ward, illustrated by Steve Jenkins (2014)
6. *A Mother for Choco,* by Keiko Kasza (1996)
7. National Geographic's Bird Pictures and Facts: www.nationalgeographic.com/animals/birds
8. National Wildlife Federation's Garden for Wildlife: www.nwf.org/Garden-for-Wildlife
9. Project FeederWatch: www.feederwatch.org
10. *Two Blue Jays,* by Anne Rockwell, illustrated by Megan Halsey (2003)

Mathematics and Science Connections

Just as the birding had inspired numerous literacy activities, it also sparked authentic, meaningful math and science investigations during activity times. A few children analyzed the information from the bird count sheet and created a pictograph to represent how many of each variety they had spotted at our feeder. This child-guided initiative led to a discussion one day about which seeds our birds preferred to eat. After doing research online about what seeds birds liked to eat, the children proposed placing different types of seeds in the feeder to observe if the number and variety of birds changed as a result.

For a number of weeks in the winter, we experimented with black oil sunflower seeds, striped sunflower seeds, mixed bird seed, and peanuts to see what worked best. I was amazed at the depth of student connection and interest. Never before had I observed the children taking such an active role in devising their own mathematical and scientific investigations. At the conclusion of our experiment, we noted that the birds ate the seeds regardless of the variety with the exception of peanuts, which only the blue jays and woodpeckers preferred. The peanuts, however, opened a new world of possibility for us as a hungry squirrel took notice of them one day in December and began to visit the feeder. As a result, we placed a separate feeder a few feet away from the bird feeder and filled it with peanuts for the squirrels each week. The children never tired of watching them climb up the feeder's pole and cling to the side. The squirrels carefully held on with one paw and used the other to open the lid and remove a peanut before scampering off to a branch to eat. I loved watching this spectacle as well.

The snowy winter also provided the perfect opportunity for us to investigate the size and shape of bird and animal tracks under the feeders. Some days we ventured outside with measuring tapes, pencils, and sketch pads to record what we found. Children delighted in seeing the variety of prints left in the snow, and they were intrigued by the markings we did not recognize. The children also examined and measured the differences in length between the tracks. We experimented by moving in the snow to observe how long our strides were when running and walking. We never did discover which other animals were eating the seeds discarded on the ground. Upon returning to the classroom, some children were inspired to paint animal tracks at the art easel, replicating what they noticed outdoors and hypothesizing as to what mysterious creatures had left them.

An Authentic, Collaborative Experience

As the weather warmed and spring arrived, children slowly lost interest in the feeders. They became more interested in bugs and butterflies, and they eagerly anticipated being able to play on the outdoor climbing equipment again. Although classroom discussions and explorations stopped focusing on the feeder activities, I was forever changed as an early childhood educator by this experience. Although the school year was a little hectic at times, I look back on it as one of my most rewarding and enjoyable. Never before in my classroom had children taken such initiative for directing their learning in so many collaborative and holistic ways. The birds sparked a natural curiosity in the children, who then spent much of their time engaged in authentic, child-centered projects and activities.

As the school year progressed, children returned time and time again to the communal experience of observing the birds, adding new personal insights to help deepen and shape their understandings. The children explored and expressed their learning collectively through literacy, mathematics, and scientific engagement. I guided the direction their inquiries took and how they researched information about birds and their behavior. Through critical thinking and collaboration, the children used different centers in the classroom to demonstrate and showcase their learning. Placing that bird feeder outside the classroom windows on that cold autumn morning was a catalyst for capturing and inspiring the children's imaginations. This empowered them to explore self-directed activities that resulted in meaningful, collaborative learning for most of the school year and hopefully instilled a love of and respect for nature for their entire lives.

CHAPTER 13

"Paper Shoes Aren't for Dancing!"

Children's Explorations of Music and Movement Through Inquiry

Coauthored by Jennifer Bombardier

For one month in the winter of 2015, the children we taught engaged in a rich, collaborative learning experience about music and movement. It began when a child's family donated three colorful, flouncy skirts to our classroom. Little did they realize those skirts would be the catalyst for the biggest self-directed inquiry the children would experience—our dance-focused Paper Shoes inquiry, a name inspired by a child's insightful comment during the experience. The project would explore the ways in which engaging in the process of art making is more powerful for young children than creating a final art product.

Amherstburg Public School is one of the first schools in Ontario, Canada, to offer Full-Day Early Learning Kindergarten, a program serving 3-, 4-, and 5-year-olds (Ontario Ministry of Education 2016). Inspired by the municipal early childhood schools of Reggio Emilia, Italy, the program uses an emergent curriculum in which children explore their interests and educators act as guides, supporting and scaffolding children's learning along the way.

We authors believe that children construct their own learning, and that they use many different ways—called many *languages,* in the Reggio Emilia experience—to communicate their thinking and understanding. Children in our classroom are encouraged to express their ideas through movement, painting, sculpture, drama, drawing, and more. These endeavors and the resulting artifacts serve as their many languages and help them internalize, reflect on, and communicate new understandings as they explore the world around them (Wien 2014). Documenting children's activities helps demonstrate to others (families, administrators, other educators, the community) that learning and play go hand in hand. By integrating children's endeavors and artifacts in our documentation practices, we honor the children's work, encourage planning and reflection through play, and share our journeys with others (ETFO 2010).

Three- to 5-year-olds never cease to amaze us. Although we have encouraged many inquiries with the children, this particular project was especially significant for several reasons. The children were extremely willing to explore and experiment through the art-making process, their interactions with one another were extended and rich, and they solved problems in the context of the play.

Inquiry is a testament to the children's rich interactions in a variety of subject areas and developmental domains. We trusted the learners in their explorations, and we supported, scaffolded, and extended the play as thoughtful and reflective educators. As a result, what began as the simple act of dancing in skirts turned into a journey that explored creative movement, the emotions inspired by music, and the representations of these feelings and ideas using child-selected languages (Wien 2014). This inquiry proved to us that when teachers are willing to look beyond the planned curriculum and honor the children as learners, children's interests can drive instruction far beyond what a rote-learning task or prepackaged theme might accomplish.

The Dancing Begins

> **When playtime starts, Anna, Levi, and María immediately notice the skirts in the recycled materials bin and eagerly try them on. Admiring themselves in the dramatic play center mirrors, they begin to sway and move to imaginary music. They giggle as they stare at their reflections, their movements becoming more complex as they add coordinated hand gestures to their rhythmic dance. María, who is rather shy and often reluctant to participate in activities, seems to lose her inhibitions when wearing the skirt. As her confidence grows, she leaps and twirls, her pleasure radiating across her face. A small crowd of children gathers, and the dancing children move to a larger space. Inspired by their audience, they improvise a series of creative movements and the crowd cheers. The spontaneous dancing lasts throughout playtime!**

Later, during sharing time, the children discussed their playtime experiences. This routine gives children an opportunity to reflect on their activities and share and expand their ideas. They had performed a dance for their peers in the skirts, and they explained how they felt while dancing. Much of their discussion focused on the dancing, and we were eager to engage the children in deeper thinking. When we asked them to describe what they thought dancing was, the children said it was "moving to music," "jumping," and "something my sister does in dance class." Their limited responses suggested the children probably had little exposure to or experience with dancing.

The next day, we were pleased to notice that the dancing continued. More children were interested in joining in and creating their own skirts. They hoped to eventually choreograph a dance that could be performed for others. We encouraged them to first design skirts by sketching their ideas on paper and then construct them using recycled materials from the art center. Putting ideas on paper helped the children conceptualize how they wanted their skirts to look and gave us an opportunity to discuss their choices.

To encourage reflection, we used various questioning techniques that required children to consider what they knew about clothing and how they wanted their skirts to function during the play ("What do you think a dance skirt looks like?," "How will you make a skirt that fits you?," "What materials will make the best dancing skirts?"). We hoped our questions would help children reflect upon their previous mathematical experiences and apply this knowledge in their problem solving. This process allowed us to discover what the children knew and to support their emergent skills, such as designing and measuring, as they ensured their skirts fit properly and stayed on while they danced. For example, the children knew that they needed to measure their waists in order to create waistbands that fit them. They had previous experience using measuring tapes from a hat inquiry, so they used these to measure one another and cut a corresponding amount of fabric for a skirt. The more the children designed, the more intricate their skirts became as they used materials from the classroom for inspiration. When their skirts did not stay clasped during dancing or did not twirl quite the way they had envisioned, the children returned to the art area, where they formed new plans and tested the skirts again and again until they had made them "perfect."

As the children modeled their skirts for one another, they discussed the similarities and differences among the designs and different ways they would approach the task next time. Children who had not been interested in constructing skirts but wanted to participate found other ways to join in. One boy lined up chairs along the carpet perimeter as seating for an audience, and another made and distributed tickets for an afternoon show. We observed that many of the children brought their strengths to the project—children who enjoyed drawing made posters for the show, and others who liked building discussed ways a stage could be constructed. These small groups of children became intensely focused on their chosen tasks, resulting in rich collaboration that the adults in the room encouraged and supported (McDonald 2018). Children independently found ways to join and belong to what was happening.

Eager to inspire children's deeper thinking within their play, we continued to use whole group learning time to introduce new materials and resources—fiction books with performance as a theme, nonfiction books about dance techniques, online information about creative movement, videos featuring different dance styles, and an assortment of music. Older children who took extracurricular dance classes visited the classroom as experts and performed for us. During playtime, the children continued to work on the project. Over the next few days they added music and built a stage using big wooden blocks. One child acted as DJ and played various clips of music for the dancers. When the children heard songs they liked, they would request them again and again, first improvising dancing to the melodies and then fine-tuning their movements to a few selected bars of music.

Next, they adapted the classroom environment to extend their play. Props for the stage, including lighting fixtures, a microphone, and curtains, were constructed in the art center; furniture was moved from the dramatic play center to the stage area to create a dressing room for the dancers; and recorded dance recitals from around the world—which the children studied and practiced along with—were played on the interactive whiteboard.

The environment became central to the children's learning. The freedom to use open-ended materials in ways that supported their interactions with one another deepened their collective understanding of what it means to dance (Leong & Bodrova 2012). For example, the children discovered that the space they had chosen for their stage wasn't suitable for maintaining the inquiry because there wasn't enough room to accommodate everyone who wanted to participate. After discussing other areas that would be more appropriate, the children agreed on a new spot in the room.

Each child brought his or her own experience, perspective, and understanding to the play. Unlike rehearsing for a school performance, there was no formal final product the children were working toward. Rather, they were engaged in these activities for the love and enjoyment of the experience. One author recalls standing on the makeshift stage in the midst of children twirling in filmy skirts to the beat of music, watching them dance with their eyes closed and their hearts open, and thinking that this was what teaching and learning are all about. This movement became the children's voice as unspoken words punctuated the classroom with each breath and each step they took. Perhaps during this experience, some children would discover a love of music and movement that would inspire them for years to come.

Connecting Music with Emotion

After a few weeks, the children continued to express an interest in dance, and we hoped to heighten their emotional connections to their experiences. We teachers had recently discussed emotions during circle time and felt this would be an interesting connection to the Paper Shoes inquiry. During group time, we introduced various styles and forms of music and asked the children to respond. The children closed their eyes and visualized the music in their minds or drew along to the melodies on large pieces of paper. They identified musical characteristics (pitch, rhythm, tempo) and talked about how the songs made them feel.

A small group of children chose to act out their drawings on the stage to express the newfound connection between music and emotions. They brainstormed more thoughtful ways to move to represent what they felt when they heard the music. Their dancing evolved from random movements improvised during play to a representation of internalized emotion. The children added facial expressions and hand gestures to help them articulate what they were feeling. When we asked them why they were doing this, they responded that they wanted their audience to know how they were feeling when dancing. The children requested the same songs over and over and worked together to choreograph movements to portray the stories behind what they heard. Although the music had no lyrics, the children had grown comfortable sharing the connections they were making with what they were hearing and using this active visualization to inspire their movements. Creating, practicing, and performing their collaborative ideas helped them realize that many points of view need to be considered in a group effort. The children also communicated their individual interpretations of the music, and together they created communal understandings through the language of dance.

During a sharing circle four weeks into our inquiry, we again asked the children to describe what dancing was. Their definitions now included "Moving my body to music I love," "Telling a story with my arms and legs," and "Like I'm floating on air." We were awed by the depth of reflective thinking these responses showed and knew that the children had expanded on their initial ideas of dancing based on collaborative explorations in their play.

Problem Solving in the Inquiry

In addition to the children's internalization of music, we were interested in the way children problem solved when they encountered a challenge or disconnect in their understanding. When some children decided to create a stage for the dancers using big wooden blocks, they negotiated a design for the stage that was agreeable to everyone and then pieced together the materials. This required them to use measurement and spatial reasoning skills as they considered questions like "How wide does the stage need to be?," "How tall does the stage need to be?," "How can the big wooden blocks be arranged so that no gaps are left between pieces?," and "How might the blocks be placed so they don't slide apart when the dancers move on top of them?"

The children also needed to continually problem solve and redesign the stage as new challenges occurred within the context of their play. When the stage wasn't big enough to include all the dancers who wanted to participate, they collaborated to redesign the space and make it even larger. When the dancers wanted to start their performance behind a curtain, the children needed to design a system for hanging a large piece of fabric and sliding it from the side of the stage. When the dancers wanted to enhance their dance skills, they brainstormed sources of information that would support their various levels of experience and expectation, and then used the resources to improve their performance. The children were thinking, learning, and communicating as they created 3-D representations of their understandings.

As they shared their work with others, questions and comments from teachers challenged them to reveal what they knew about dance and what they needed to find out. This process continued until the children had refined their thinking, deepened their understanding, and become "experts." When the children felt they had met their goals by indicating their satisfaction with their participation in the play, their interests evolved and they moved on to other more complex areas of exploration—more elaborate movements to music, more intricate costumes, and a grander stage.

To support the children's interactions, we invited them to engage in further conversations with us during playtime. We hoped to help enhance their emerging understandings and ask questions to provoke deeper thinking about their collaborative work. During whole group time, the children continued to share what they had created and explored in their play. As they discussed why they had created their skirts in the manner they had, choreographed movements in a particular series, or reinvented materials to make objects needed to support their play, they engaged in deep reflective thinking. They experienced pride in their accomplishments as they celebrated with peers. The excitement they had for their endeavors was contagious, and we often noticed other children who weren't initially involved in the play join in after hearing their peers speak.

Sharing time helped us discover what questions the children still had and how we could support their efforts to continue refining their understandings of what it meant to dance. A particular concern of the children was that the shoes created from paper didn't last through a session of dancing, and this was frustrating. As one child emphatically shared as she examined her torn shoes, "Paper shoes aren't for dancing!" We encouraged the children to reconsider the materials and design of the shoes and think about how they could improve them. We wanted the children to understand why each style of dance has its own shoe, so we brought in a variety of footwear, including tap shoes, ballet slippers, and jazz flats, for them to examine. We encouraged the children to explore how the shoes were constructed and to observe how their feet felt when dancing in them. Using online videos, we helped the children understand what materials were used to make the different shoes. The children were fascinated with this new knowledge and brainstormed how to create more durable shoes.

> **Eric is intrigued with the sound that tap shoes make on the tiled floor, but disappointed that the shoes don't fit his feet. He decides to create tap shoes for himself, using his running shoes as a guide. As Eric studies the tap shoes, he discovers a metal piece that makes the tap sound when it touches the floor. He finds lids from frozen juice containers in the art center and attaches them to the bottoms of his shoes with masking tape. The smile on Eric's face and the sound of tapping on the tiled floor lets us know that his plan was successful! Other children gather around to share in his accomplishment and ask how to make their own tap shoes. Eric enjoys his role as an expert, coaching his peers as they attempt to replicate his idea using their own shoes.**

As the children constructed and tested, we asked questions to elicit explanations for their design choices and ideas about how the designs could be improved. Through this cycle of planning, experimentation, and reflection, the children formed a more complex understanding about how shoes are constructed. They felt safe taking risks and sharing their learning with others.

Representing Feelings and Ideas in Their Own Languages

Throughout this inquiry, the children shared their understandings with others through their performances and 2- and 3-D representations. Some of these became props and were interwoven in their play, such as the newly designed dance shoes and stage, while others were artistic expressions of dance displayed throughout the room so that the environment could support and extend the play. The children expressed their ideas in various ways—painting pictures of themselves dancing, sculpting dancers using playdough, dramatizing dance performances with puppets, and writing about personal experiences in journals.

As we examined this collection of representations, we noticed a level of detail that suggested the children were internalizing their interpretations of what dance was. The paintings were vibrant with color and detail, the sculptures were formed so they appeared to be leaping through the air, the puppet shows reflected life in the classroom, and the journal entries shared personal narratives of the children's experiences. We used photography and written teacher narrative to add our

interpretations of what the children had accomplished. Throughout the project we continually took photos and made videos of the children dancing, adding our interpretations in the form of text to accompany the recordings. This evidence of the children's thinking was posted throughout the classroom.

Through the children's many languages—painting, sculpting, puppetry, role-playing, and drawing—a complex, communally developed representation of the children's understanding of dance emerged. Families were invited to observe and discuss this documentation with their children in the classroom. Sharing the experience encouraged the children to reflect on their learning and inspired deeper thinking. Together with their teachers and families, they celebrated the richness of their play. Furthermore, it gave the teachers an opportunity to advocate for the importance of play-based learning, demonstrating how vital it is to children's development.

Conclusion

When offered a wide array of materials and opportunities to explore them, children construct their own meanings and interpretations about their world. By emphasizing the process and not the product, the children participating in the Paper Shoes inquiry were free to explore and experiment. Through a variety of authentic experiences driven by each child's personal interests and strengths, the children socially constructed their own knowledge and understanding (Leong & Bodrova 2012; McDonald 2018; Wien 2014). When teachers value and promote the arts, children use them as regular methods of communication in the classroom and beyond. The joy born from children expressing their emotions through movement may inspire them to keep dancing. Long after our dance inquiry had ended, we occasionally found a lone child, still wearing her tattered skirt, twirling in front of mirrors in the corner of the room, lost in a performance all her own.

About the Coauthor

Jennifer Bombardier, BA, RECE (Registered Early Childhood Educator), works in the Full-Day Early Learning Kindergarten Program with the Greater Essex County District School Board in Windsor, Ontario, Canada. Her passion for the early years focuses on inquiry-based learning and embraces emergent philosophy and the arts.

CHAPTER 14

Mitten Measurement

Inspiring Rich Math Inquiry Through Children's Books

> "That's not possible in real life! There's *no way* all those animals would fit in there!"
>
> I am reading *The Mitten,* by Jan Brett, aloud to my class of 5- and 6-year-olds during circle time. Recently, the children expressed curiosity about how animals live during the winter. As part of our ongoing explorations into this interest, I thought reading this book together would be a fun complement. Toward the end of the story, many animals including a hare, a hedgehog, and a bear squeeze inside a woolen mitten lost in the snow to stay warm. It is at this point that 5-year-old Harlow spontaneously calls out what she thinks.
>
> "Yeah," Harlow continues, "no way. I mean, how can a real, live bear fit into a teeny tiny mitten?"
>
> Other children nod their heads in agreement, beginning to voice their opinions too.
>
> "It wouldn't even be big enough for a rabbit," points out Aroon.
>
> "I don't think any animal would want to be next to the hedgehog," Nanurjuk says doubtfully. "They would need space between them."
>
> "When I grew bigger, my hands didn't fit into my old mittens anymore," says Arthur.
>
> Nora adds, "If all those animals were in one mitten, it would need to be bigger than our classroom!"
>
> In this moment, it's clear that a new inquiry is sparking. I am interested in how the children's previous experiences with measurement are emerging in this conversation. Marking our place in the book, I set it aside for now and ask the children, "So exactly how big do you think the mitten would have to be for all of the animals to fit inside of it in real life? How might we calculate the size?"

Children's wonderings about the world and the sparks of new curiosity that naturally arise as they research, observe, and communicate about a topic of interest often cultivate rich opportunities for complex, joyful math learning. As part of my own interest in math, I am always seeking opportunities in the classroom that emphasize authentic math situations and focus specifically on developing children's reasoning and problem-solving skills.

When the children expressed their disbelief that the mitten in the story could realistically contain all the animals, I was immediately drawn to the mathematical potential of their observations and comments. There were many questions I considered as I imagined how best to harness and facilitate

this emerging inquiry. What collective knowledge about and experiences with measurement did the children already have? How could I tap into their existing interest in exploring the size the mitten would have to be to realistically retell this story? Could the children rationalize their opinions and make predictions using relevant math terminology? How could I effectively support and honor their mathematical explorations and help deepen their thinking?

Using Children's Books as a Bridge to Math Learning

There are multiple benefits to using rich children's books as entry points for exploring math concepts. Stories offer a context that children find relatable and can emotionally connect to. They also feature a meaningful idea or problem scenario that children can explore, increasing their motivation as well as promoting their critical thinking and problem-solving skills (Flevares & Schiff 2014). Because children are engaged and emotionally invested in the scenario of a story, they are inspired to help in some way, and this inspiration feeds into complex, hands-on, inquiry-based explorations. With your guidance, these explorations can be mathematized.

As I listened to the lively discussion that followed our reading of *The Mitten,* I was impressed with the math vocabulary that the children were incorporating in their conversation and how they constructed logical arguments based on their past measurement experiences. One of the first things the children determined was that the mitten would have to be very large—*much* larger (comparison) than the one portrayed in the book—to fit all of the animals. They also sympathized with the grandmother in the story who had knit the original mitten that the animals stretched out.

The children agreed that, in reality, it would be very time consuming to knit a mitten big enough to house all of the animals. If the grandmother were to knit one, the children recognized that she would have to know exactly how big that mitten needed to be (attention to precision)—if it was too small, the animals wouldn't fit; if it was too big, it would waste yarn and her time.

Intrigued with the complexity behind their thinking and interested in driving their math curiosity forward, I challenged them to consider how they might create a mitten in our classroom the exact size needed to fit all of the animals from the story: a mole, a rabbit, a hedgehog, an owl, a badger, a fox, a bear, and a mouse. To refine their inquiry, first we analyzed what the children knew, what they didn't, and what problem they hoped to solve. Then we narrowed the scope of their exploration to a single question: What size does the mitten need to be to hold all of the animals from the story?

Finding and Representing the Data

In order to answer their question, the children determined that they needed to research the exact size of each animal. They agreed that once they knew how big each animal was, they could create a realistically drawn representation for each one. The thinking behind their decision to create a sketch as part of their inquiry was connected to their previous work with math models, during which they drew blueprints before creating structures using large, wooden blocks. As a starting point for gathering this information, they suggested finding books about animals at the school library and researching animals online. In small groups, we visited the library, where we gathered and looked through a large collection of animal books. In pairs, children also spent time on the library computers searching for important information that would help them identify and record the size of each animal.

Once the children found the information they were looking for, their next step was to interpret the data. As we researched the animals, I introduced terms like *length, width, height,* and *weight* as we considered each animal's size. Due to the children's limited experience with standard units of measurement (inches, pounds), I helped them convert the size of each animal into a nonstandard unit of measurement that was meaningful for them—connecting cubes ("The mouse is 2 connecting cubes tall and 4 cubes wide"). Using connecting cubes, the children then estimated how large a sheet of paper they would need for each animal and then created an outline of the animal on the paper. We discussed how these drawings were two-dimensional (2-D) approximations of the size of each animal. The children quickly discovered that the bear was the largest, taking up almost a third of our carpet when outlined using cubes!

After creating a connecting cube perimeter for each animal, the children traced their shapes on the paper and then cut them out. This process took a few days to complete, and it was something the children returned to eagerly during free choice time. As each animal was created, it was placed on a large table. The children were amazed to compare each animal—many of them had never seen these animals up close or at all in real life—and their conversations were excited and filled with rich math discussions regarding their observations ("Look how much taller the owl is than the rabbit!" and "If I lie down, I think I'm as long as the fox"). The paper animals became math tools that allowed the children to experiment with different concepts as they explored. They lined the animals up from smallest to largest, sorted and grouped them using different attributes (number of legs), and compared their sizes to different objects in the classroom ("This mouse would fit in my lunchbox!" and "I think the badger could sit on our chair").

A Mitten Built for Eight

The next step in our inquiry process was to calculate the actual size of the large mitten needed to hold all of the animals. After listening to the children discuss various strategies for how to go about doing this, I entered their conversation, eager to continue embedding mathematical thinking into their planning and ready to answer any questions they might have. I suggested that we refer to the book to see if the animals were side by side or on top of each other after crawling inside the mitten. The illustrations weren't clear, so the children decided to place each animal side by side in order to be as accurate as possible. After much debate, the children planned to unroll a very large piece of butcher paper next to the row of animals in order to estimate how long the mitten needed to be. In a similar fashion to how we created each animal, we used connecting cubes to determine how wide the butcher paper should be to accommodate even the widest animal of the bunch, extending the paper's width with additional paper and tape. We then drew a large mitten on the paper, traced a second copy of it onto another piece of butcher paper, and worked carefully to cut out both versions, which were then stapled together to make a paper mitten. Again, this took several days to complete.

Persevering through a few mistakes, the children realized that pieces of butcher paper cut too short could be taped together in order to make the mitten long enough and pieces that were too long could be trimmed to the correct length. They also realized that to create a pocket in which to place the animals, both halves of the mitten needed to match so they could be secured together. Once complete, the mitten was almost the size of our carpet! The children were surprised at just how big it was and thrilled with the progression of their work. I invited them to bring the large mitten back to our sharing circle later in the day in order to talk about the problem solving they used to calculate the size of the mitten and the steps taken to construct such a large object. This was especially important as not all children had participated in the experience and I wanted them to become familiar with the problem-solving strategies their peers had used in the math inquiry.

Finally, it was important that we test our design. After the mitten was presented to the whole group and the children were satisfied it was created accurately, they suggested retelling the story and placing each animal within the paper mitten in sequence. We gathered around our circle time carpet with the mitten in the middle. As I read the story aloud to the class page by page, the children took turns placing the paper animals into the mitten. They were thrilled—in the end, all the animals fit into the mitten and the exclamations and conversations that erupted were rich with mathematical observations and connections. They compared the size of the mitten to the original in the text and marveled at the complexity of their creation. They estimated how many balls of yarn might be needed to knit a mitten the size of their paper creation and guessed how long the process of knitting it might take. One child shared that his grandma was a skilled knitter and that he would ask her next time he visited and report back to our class. I wondered aloud if his grandmother might be willing to visit our class in order to demonstrate her knitting skills, helping us to connect our mitten inquiry to the outside community. I anticipated the additional math wonderings this experience might inspire.

Once the retelling was complete, the children had another question they wanted to explore: Would a mitten that was large enough for all of those animals hold 14 kindergarten children? They were quite eager to find out! They gently removed the paper animals and placed them aside. Then, one by one, they wriggled and squeezed inside the mitten, making room for everyone to join in! Unfortunately, their paper mitten was not built strong enough for a group of giggly, wiggly kindergartners. The children burst through the mitten with laughter

and talk about how much fun it would be to have a mitten that was large enough to play inside of. Many made connections to camping experiences and shared stories of snuggling inside sleeping bags and tents. Soon, the conversation shifted toward planning a campsite in the dramatic play center later that day. With that, another spark had ignited and the children moved forward in their explorations, eager to construct a different kind of structure within which they could hide and play.

Conclusion

The most effective math learning experiences are those that help children explore real-life problems and connect them personally to a question to be explored. Children's books are rich with mathematical potential as children delve deeply into texts and use interesting story lines and illustrations as inspiration for math learning. Educators who share high-quality children's books and are attuned to children's discussions and wonderings about those stories can support authentic, meaningful math in emergent classrooms. Cultivating a safe and supportive space where critical thinking is embedded throughout and the mathematical relevance is intentionally explored empowers children to bring their ideas to life in new and innovative ways!

CHAPTER 15

Counting Kindness

A Food Drive Inspires Rich Mathematical Thinking

Three 5-year-olds, clipboards in hand, stand next to a table holding food donations. "We have 26," Jorge says, and gestures toward the overflowing pile.

"Yes, 26," Sam repeats, while double counting the circles he has recorded on his clipboard.

"I have that, too," Maalai chimes in.

"So, 26 here and 16 in the hallway. That makes . . . 42." Jorge verifies his total and writes 42 on a sign posted next to the food donations in the hallway.

"How did you know that?" I ask him.

"Easy! Six plus 6 is 12, and then I added that to the other numbers, 20 and 10," Jorge says proudly.

I nod, amazed at this kindergartner's ability to mentally add these two-digit numbers.

How did a 5-year-old come up with this mental computation? Did the kindergartners in our program learn this strategy in class? Or did Jorge figure it out himself? The latter. Jorge's deep involvement in the class project prompted the mathematical thinking necessary to quickly compute important sums.

In the program where I teach, we embrace inquiry. Each day is new and exciting as teachers follow the children's lead in crafting authentic math inquiries in play-based learning centers. We have clear learning objectives, and we strive to achieve them through an emergent curriculum in which we use children's interests as a spark to motivate and engage. In this chapter, I discuss our approach to developing children's mathematical knowledge and skills.

Emergent Math Curriculum

In emergent programs, educators aim to embed rich math resources and learning opportunities throughout the room—not just in a defined math center—in order to encourage organic exploration. For example, we add dominoes and dice to loose parts play, use connecting cubes and links in the building area, and take ten frames outdoors. Children investigate and find solutions to questions or problems generated in the social interactions they have during playtime. By carefully observing children's play, teachers can bring children's inquiry-based topics and authentic queries to the whole group for extensive discussion and investigation, specifically highlighting math and computation, when applicable. Because children are highly invested in solving play-based problems, they are more likely to work together to brainstorm various strategies for arriving at solutions.

An emergent kindergarten program creates natural opportunities for purposeful computation by

- Providing children with the freedom to explore self-directed areas of interest
- Inviting children into a thoughtful space with interesting and challenging materials for them to use independently and with teacher guidance
- Emphasizing a child-centered problem-solving approach when difficulties arise
- Honoring children's questions and providing guidance when needed
- Providing multiple opportunities for children to safely take risks and make mistakes

Meeting Expectations

One concern many educators face when following an emergent curriculum is whether they are successfully covering expectations in the classroom. In Ontario, teachers are required to follow the Full-Day Early Learning Kindergarten Program, which outlines numerous big ideas in each subject area that must be implemented (Ontario Ministry of Education 2016). Many educators, however, are inspired by the emergent model and seek to follow children's leads, creatively integrating areas of the curriculum to effectively meet the standards.

In addition, many programs have improvement plans focused on a particular subject or area of need that are implemented on a year-by-year basis. Interested in establishing an improvement plan focusing on children's mental math abilities, my colleagues and I read Sherry Parrish's work regarding number talks (2010, 2011). Parrish, a mathematics educator and consultant,

recommends engaging children in regular, short conversations about purposeful problems. These problems and conversations are intentionally crafted to draw children into specific computational situations. When we observe children in play, we look for—and explicitly draw children's attention to—mathematical opportunities. For example, a teacher may say, "I see you are sharing the cubes with Joaquin. You two have equal amounts, which makes it fair, because you both have the same number of cubes."

Parrish dissuades the use of paper-and-pencil activities before first grade. Worksheets usually have a "right" answer and do not account for differences in interests and current abilities. "When children approach problems without paper and pencil, they are encouraged to rely on what they know and understand about the numbers and how they are interrelated. Mental computation encourages them to be efficient with the numbers to avoid holding numerous quantities in their heads" (Parrish 2011, 204). With our emergent approach and math opportunities spread throughout the learning environment, we saw Parrish's number talks as a great way to amplify children's play-inspired learning through child and teacher math conversations.

I was fortunate to experience a powerful example of emergent curriculum when the children in my classroom became fascinated by a program-wide food drive. We teachers were excited to observe what they would do to overcome the challenges that arose—and we knew we would have plenty of opportunities to support the children's math development with authentic problem solving.

Food Drive-Motivated Math Learning

Our program conducts an annual food drive. Teachers introduce it to the children by discussing the value of helping others in the community. They actively promote the food drive with families by, for example, advertising it in class blogs and speaking with families at drop-off time. For my kindergartners, the food drive always creates many authentic opportunities for math learning, but one year the children were particularly inspired: The class with the highest number of items donated would win a pizza party.

Each morning, as more food items were donated, the children eagerly calculated the total number of items and compared their total with other classes' collections. Day to day, we never knew how many items would be donated. The children experimented with a variety of strategies and tools for doing their calculations and storing the donations. As a result of the children's problem solving, teachers noticed richer math talk emerging in other areas of the classroom and saw children applying number sense strategies in learning centers, resulting in a more robust and accurate exploration.

Geometry Terms

As food was donated, it began to fill the hallway, and the children became creative in displaying the donated items. They used their sense of measurement and spatial relationships to stack cans high enough to be out of the way yet not tall enough to topple over. They were challenged by how to efficiently fit various sizes and shapes of packages into boxes. The children wanted to have all items on display but were limited by the available space. If the food was not organized properly, it could become a safety issue—the hallway would be impassable. What a wonderfully motivating opportunity for real-life problem solving!

The children approached organizing the food as they would completing a jigsaw puzzle—the shapes of the cans and boxes required them to be fitted together in certain ways to maximize the available space. As they worked with the items, the children automatically named the shapes they noticed—"This box looks like a rectangle," "This can has two circles on it"—giving educators natural opportunities to discuss the differences between 2- and 3-D shapes and introduce the names of new shapes (such as rectangular prism and cylinder). Helping children identify shapes in the world around them and challenging them to manipulate these shapes in various ways established a solid foundation for future geometric work (Moss et al. 2016). As the children became comfortable with the terminology, they used the words more frequently. Sorting the items by observed characteristics, such as shape and size, the children placed cylinders in one pile and rectangular prisms in another. Irregularly shaped items, damaged goods, and items with interesting packaging went into their own collection; the children recognized that these were not common shapes and did not fit any of the sorting characteristics they were using.

Number Sense

As donations accumulated, the children could no longer rely on familiar strategies to calculate the total number of items they had collected. Early on, they could rote count the total number of items already collected and add the new items to the collection to find the total ("We have 12 in the hallway, so 12 plus five more is 13, 14, 15, 16, 17. Now we have 17"). By the middle of the food drive, the numbers were much bigger than the children were used to working with, and the rote counting strategy no longer worked well. My colleagues and I encouraged the children to brainstorm how they would approach this new challenge. We felt that encouraging logical problem-solving strategies, and not focusing on the speed or accuracy of the solutions, would encourage more children to participate in the experience (Parrish 2010).

Working with large numbers requires children to have solid number sense, which we developed through many experiences with subitizing and place value.

Subitizing

We had spent a great deal of time before the food drive (which was done shortly before winter break) embedding the skill of subitizing in meaningful contexts. Subitizing is the ability to instantly recognize a quantity of objects without counting them (the pips on the face of a die equal five). Subitizing is an important foundation for successful number sense because it helps children develop a solid grasp of what numbers actually represent. It is like taking numerical snapshots (Boaler 2016). Children are shown an image, such as a collection of dots, and take a mental photograph of the collection, helping them identify how many dots are included in the image. To reinforce this skill, we challenged children to play several games using dominoes and dice. They became quite proficient at recognizing small quantities in groups and in mentally adding small groups of numbers. To help them transfer this skill to food drive calculations, we encouraged the children to group the items for quick counting. Many children could count by twos and fives with ease, so they used this approach to find a total for the first few days of the drive, when the number of donated items was relatively small.

Internalizing Place Value

The number of food items continued to grow, and it was no longer efficient to count by ones, twos, or fives. Earlier in the year, we had explored five frames and ten frames, and some of the children realized that ten frames could be used to organize and total the food donations. We photocopied large ten frames, and the children placed items on them, creating multiple groups of 10. They then quickly counted by 10s to calculate their totals and added the remaining numbers with ease (10 full ten frames and a ten frame with only two items totaled 102 items). Each day the children wrote the new total on a sign posted alongside the hallway display.

Children need many opportunities to practice making and organizing groups of 10 to construct a solid understanding of place value (Parrish 2010). Wanting to gauge their progress and make their math learning more explicit, we asked the children whether they could write down an addition sentence using numbers to represent their thinking about the donations in the ten frames. We modeled how an addition sentence could be recorded vertically and the numbers placed in columns. As the children gained confidence, they built a more complex understanding of place value and realized they could add numbers by carrying quantities from the ones and tens columns to efficiently arrive at an exact count of the food items each day. We believe that their ongoing focus on subitizing and the regular use of ten frames and other math tools (dice, dominoes, playing cards) helped to support children's number recognition, leading to a stronger ability to compute basic numbers and to internalize an initial sense of place value.

Of course, some children became more engrossed in the food drive—and winning the pizza party—than others, but even children who did not participate in the daily donation count were observed using math in new and exciting ways. Because encouraging the children to frequently and freely engage in math talk was part of our improvement plan, my colleagues and I took great notice whenever it happened. Some of the children's documented math talk included using blank ten frames in their play (organizing toy cars on ten frames and using these as parking lots to see how many cars fit into a garage constructed from blocks), incorporating more numbers into their independent work (recording numbers in their journal writing), and using math language more frequently in their interactions (using the words *more* and *less* to describe collections).

Representing Numbers in Different Ways

Throughout the food drive, we helped children understand that a number can look many different ways but still equal the same amount (15 can be represented by three groups of five tallies, a picture of three hands totaling 15 fingers, or one filled ten frame and one half-filled ten frame). We invited children to represent the numbers of food donations on a dry erase board using these strategies (which allowed us to reinforce our explorations of five and ten frames from earlier in the year). Figuring out how to write a very large three-digit number in words and pictures required a lot of discussion; it encouraged higher level thinking as the children explored various ways of showing others how many items had been collected. As one enthusiastic child said, "We need everyone to know what we've done!"

The Big Count

For days, the children had counted down to the final day of the food drive. My colleagues and I anticipated that the final food donation count for our kindergarten classroom would be over 400 items. While the children had calculated daily totals as items came into the classroom, this grand total count would be their most challenging work with larger numbers. To include as many children as possible in representing the grand total, and to support all of the children in working with such large numbers, we felt that using ten frames and a hundreds chart could organize the large group of food items into manageable smaller groups and ensure an accurate count. The visual aids and the act of manipulating the items into tangible collections would help differentiate the task so more children could participate.

After brainstorming a plan with the children at morning circle, we decided to use a very large hundreds carpet to help us group our food. This carpet had 100 equal-size squares, each large enough to hold an actual item. The children quickly worked to fill the squares. As they placed the items and filled the rows, they orally calculated the totals: "That's one row of 10," "Now we have four rows, so we're at 40," "Sixty now, so we have 40 to go." Once the grid was full, the children moved the group of 100 items into a collection off the carpet. They repeated the process four times.

With fewer than 100 items left, the children knew that they could still use the hundreds grid one final time to count the remaining items (counting by ones) or they could use ten frames and get an efficient count more quickly. Using six ten frames, they filled them with items until five were full and the final ten frame had four items on it. That meant they had collected 454 items for the food drive—an impressive total and what we all hoped would be enough to win the coveted pizza party! The only thing left to do was wait for the office to announce the winning class. Sadly, it was not our class, but the children's hard work was acknowledged. One parent was so impressed with the children's efforts, she contacted a local radio station, which awarded the kindergartners Class of the Week, complete with a pizza party.

Conclusion

Infusing number talks into the food drive was clearly a winning strategy. One morning not long after the food drive, the class was discussing ways to compose the number 15. One child shared that the number 15 was like three hands: Three groups of five fingers equal 15. He said this was like multiplying 3 times 5 to total 15 ($3 \times 5 = 15$). Another child added that if he put his hand together with the other three, then it would be four groups of five, which would total 20 ($4 \times 5 = 20$). What challenges might emerge from this initial discovery of multiplication, and how would the children use this knowledge to further their learning and classroom investigations? We were excited to find out!

We learned that when rich, everyday math talk is encouraged and supported, there are endless opportunities for real-life problem solving (Parrish 2010). Numbers and calculations become meaningful and accessible to the children as they are driven to solve a problem. A major goal of our program's improvement plan was for the children to see themselves as capable mathematicians able to take on the challenges of math without the fear of failure. Through an emergent curriculum that infused real-life math into the children's experiences, we found that it is possible to cultivate an environment that supports and encourages differentiated approaches for children, allowing everyone to take on new and exciting math challenges.

Appendix

Reproducible Resources for Families

The joy of exploring and playing with math can continue at home! The resources on the following pages focus on some of the information and strategies covered in this book, taking key ideas and turning them into one-page handouts that you can photocopy and share with children's families.

Five Frame Subitizing

A Family Math Game

Subitizing is the ability to see and recognize a small number of objects and know instantly how many there are without counting. Subitizing tells you what number you roll on a six-sided die—most adults no longer have to count the dots after playing board games for a while. It's an important foundational skill for children's development of number sense. Try this activity with your child to expose them to subitizing!

Materials

A five frame (helps children show quantities 1 through 5)

Counters (shells, coins, etc.)

How to Play

Step 1: Start the game by showing your child the quantities 1 through 5 in the five frame. Put one counter in the first box and say "one." Now, it's your child's turn to show one. Next, put a counter in each of the first two boxes and say "two." Take turns to five.

Step 2: Play the game again, this time with your child going first.

Don't worry if your child doesn't catch on to subitizing right away. The important thing is to have fun and enjoy math together.

For more fun activities to do with your child at home, visit www.mrsmclennan.blogspot.com.

This message is meant to be photocopied and sent home with children as a resource for families.

2

Explore and Play with Math at Home!

While your preschooler isn't yet ready to memorize multiplication tables, he or she is a natural mathematician. Math is all around us, and every day in the classroom, children are exploring math concepts they will use when they move on to kindergarten and beyond. Here are a few ideas you can try at home to help build your child's growing math skills.

1. **Offer containers filled with small treasures.** Bottle caps, buttons, shells, beads, pieces of ribbon, pinecones, acorns, and similar items are great math learning tools. Your child will naturally sort them by size, color, and shape; count and compare collections; and talk about what they are doing and why—especially if a grown-up joins in. *I saw you examine each button before placing some of them in the blue bowl. What were you looking at?*

2. **Talk about math.** Include math talk during your daily routines—cooking, playing at the park, and at bedtime. *Our family has five people eating dinner. How many ears of corn should we shuck? How many times do you want me to push you in the swing? We can read three books together before turning out the light. You can choose three books from the shelf.*

3. **Measure things.** Your child enjoys using measuring tools, both standard ones (like rulers and measuring tapes) and more creative items (like shoes and paper clips). Ask questions and make comments that invite your child to measure something. *How wide is your bed? I wonder how tall our dog is. How many shoes long is the carpet?*

4. **Build together.** Make buildings and other creations with wooden blocks, LEGO bricks, and even recycled items like shoe boxes and paper towel rolls. Try using masking tape to hold the structures together. As you build, talk about shape, size, width, and height. Then get out the tools and take some measurements. *How did you make the building so high? How wide is the bottom of the spaceship?*

For more tips and ideas about how you can explore math concepts at home with your child, visit www.robmadell.com and NAEYC.org/our-work/families/math.

This message is meant to be photocopied and sent home with children as a resource for families.

References

Boaler, J. 2016. *Mathematical Mindsets: Understanding Students' Potential Through Creative Math, Inspiring Messages, and Innovative Teaching*. San Francisco: Jossey-Bass.

Daly, L., & M. Beloglovsky. 2015. *Loose Parts: Inspiring Play in Young Children*. St. Paul, MN: Redleaf Press.

Dietze, B., & D. Kashin. 2018. *Playing and Learning in Early Childhood Education*. 2nd ed. North York, ON: Pearson Canada.

ETFO (Elementary Teachers' Federation of Ontario). 2010. *Thinking It Through: Teaching and Learning in the Kindergarten Classroom*. Toronto: ETFO.

Flevares, L.M., & J.R. Schiff. 2014. "Learning Mathematics in Two Dimensions: A Review and Look Ahead at Teaching and Learning Early Childhood Mathematics with Children's Literature." *Frontiers in Psychology* 5 (459). doi:10.3389/fpsyg.2014.00459.

Flynn, M. 2017. *Beyond Answers: Exploring Mathematical Practices with Young Children*. Portland, NH: Stenhouse Publishers.

Grossman, S. n.d. "The Worksheet Dilemma: Benefits of Play-Based Curricula." Earlychildhood NEWS, accessed February 20, 2020. www.earlychildhoodnews.com/earlychildhood/article_view.aspx?ArticleID=134.

Heick, T. 2019. "4 Phases of Inquiry-Based Learning: A Guide for Teachers." *TeachThought*, November 5. www.teachthought.com/pedagogy/4-phases-inquiry-based-learning-guide-teachers.

Leong, D.J., & E. Bodrova. 2012. "Assessing and Scaffolding Make-Believe Play." *Young Children* 67 (1): 28–34.

McDonald, P. 2018. "Observing, Planning, Guiding: How an Intentional Teacher Meets Standards Through Play." *Young Children* 73 (1): 31–35.

Moss, J., C.D. Bruce, B. Caswell, T. Flynn, & Z. Hawes. 2016. *Taking Shape: Activities to Develop Geometric and Spatial Thinking, Grades K–2*. Toronto: Pearson Canada.

NAEYC & NCTM (National Council of Teachers of Mathematics). 2002. Updated 2010. "Early Childhood Mathematics: Promoting Good Beginnings." Joint position statement. Washington, DC: NAEYC. www.naeyc.org/sites/default/files/globally-shared/downloads/PDFs/resources/position-statements/psmath.pdf.

Ontario Ministry of Education. 2016. *The Kindergarten Program, 2016*. Toronto: Queen's Printer for Ontario. www.edu.gov.on.ca/eng/curriculum/elementary/kindergarten.html.

Parrish, S. 2010. *Number Talks: Helping Children Build Mental Math and Computation Strategies, Grades K–5*. Sausalito, CA: Math Solutions.

Parrish, S.D. 2011. "Number Talks Build Numerical Reasoning." *Teaching Children Mathematics* 18 (3): 198–206.

Pecaski McLennan, D. 2017. "Books that Introduce Coding." *Teaching Young Children* 10 (3): 22–23.

Umaschi Bers, M. 2018. *Coding as a Playground: Programming and Computational Thinking in the Early Childhood Classroom*. New York: Routledge.

Wien, C.A. 2004. "From Policing to Participation: Overturning the Rules and Creating Amiable Classrooms." *Young Children* 59 (1): 34–40.

Wien, C.A. 2014. *The Power of Emergent Curriculum: Stories from Early Childhood Settings*. Washington, DC: NAEYC.

Wurm, J.P. 2005. *Working in the Reggio Way: A Beginner's Guide for American Teachers*. St. Paul, MN: Redleaf Press.

Resources

Articles

Geist, E. 2016. "Let's Make a Map: The Developmental Stages of Children's Mapmaking." *Young Children* 71 (2): 50–55.

McClure, E. 2017. "More than a Foundation: Young Children Are Capable STEM Learners." *Young Children* 72 (5): 83–89.

Wien, C.A., & D. Halls. 2018. "'Is There a Chick in There?' Kindergartners' Changing Thoughts on Life in an Egg." *Young Children* 73 (1): 6–14.

Books

Aspinall, B. 2017. *Code Breaker: Increase Creativity, Remix Assessment, and Develop a Class of Coder Ninjas!* San Diego: Dave Burgess Consulting.

Fleming, L. 2015. *Worlds of Making: Best Practices for Establishing a Makerspace for Your School.* Thousand Oaks, CA: Corwin.

Heroman, C. 2017. *Making and Tinkering with STEM: Solving Design Challenges with Young Children.* Washington, DC: NAEYC.

Hynes-Berry, M., & L. Grandau. 2019. *Where's the Math? Book, Games, and Routines to Spark Children's Thinking.* Washington, DC: NAEYC.

NAEYC. 2015. *Exploring Math and Science in Preschool.* Washington, DC: NAEYC.

Novakowski, J. 2015. *Reggio-Inspired Mathematics: A Professional Inquiry Project in the Richmond School District.* Richmond, BC: Richmond School District.

Online

Kaldor, T. 2017. "The T in STEM: Creating Play-Based Experiences That Support Children's Learning of Coding and Higher Order Thinking." *NAEYC* (blog), February 1. www.naeyc.org/resources/blog/creating-play-based-experiences.

NAEYC & Fred Rogers Center for Early Learning and Children's Media. 2012. "Technology and Interactive Media as Tools in Early Childhood Programs Serving Children from Birth Through Age 8." Joint position statement. Washington, DC: NAEYC. www.naeyc.org/sites/default/files/globally-shared/downloads/PDFs/resources/topics/PS_technology_WEB.pdf.

Pecaski McLennan, D. *Joyful Learning in the Early Years* (blog). www.mrsmclennan.blogspot.com.

Acknowledgments

This book is possible because of the tremendous support, encouragement, and love I receive from many special people in my personal and professional life.

To my parents, Antoinette and Ernest Pecaski, thank you for your unwavering support over the years. Your exemplary careers as educators helped shape me into the teacher I am today.

To my husband, Trevor, I love you more than words can express. I am so thankful for having you in my life. To my three children, Cadence, Caleb, and Quinn, thank you for filling my life with love and joy.

To the amazing children, families, administrators, and colleagues with whom I've had the pleasure of working over the years, thank you! You inspire me to be the best educator I can be each day. I am a proud public school educator in Ontario's groundbreaking full-day kindergarten program, which has tremendously benefitted the youngest learners in our province.

To the many editors and designers with whom I've had the pleasure of working over the years at NAEYC, thank you for believing in my work and helping me share it with a larger audience! Countless conversations, revisions, and reflections helped me to improve my pedagogy and practice, benefitting hundreds of children and families over the years. A special thank you to Rossella Procopio and Gillian Frank for their help with this book.

To my grandmothers, Carmela DeThomasis and Mary Pecaski, thank you for instilling in me a love of stories and an appreciation for the power of education. Although you are no longer with me, I feel your presence in my life each and every day.

About the Author

Deanna Pecaski McLennan, PhD, is a passionate early childhood educator, researcher, and writer from Amherstburg, Ontario, Canada, who has spent over 20 years working with young children. As a teacher in Ontario's full-day kindergarten program (equivalent to US prekindergarten) and a university instructor, she devotes her research and practice to exploring the potential for rich mathematics learning through playful inquiry and exploration. She appreciates looking at life through a mathematical lens and recognizes opportunities for authentic, complex engagement in all experiences children have. She is a passionate advocate who feels deeply that the investments we make in the lives of young children benefit our society tremendously over time.

Deanna is also a mom to three children and knows firsthand the importance of supporting and improving their confidence, fluency, and accuracy in mathematics. She considers herself a lifelong learner and has spent the last several years transforming her classroom into a safe, supportive space where children can take risks without the use of rigid programming, rote practice, or worksheets. Deanna believes that educators benefit from sharing their journeys with one another and actively uses reflective writing, technology, and social media to connect with others from around the world. Part of her journey has involved writing extensively for NAEYC and publishing articles in the organization's magazine, *Teaching Young Children*, and peer-reviewed journal, *Young Children*. Deanna uses her writing and large social media following to provoke conversations about the role math plays in emergent programs and how educators can cultivate children's mathematical mindset while still honoring their interests, strengths, and needs. You can learn more by visiting her website at www.mrsmclennan.blogspot.com or following her on Twitter and Instagram @McLennan1977.

Discover NAEYC!

The National Association for the Education of Young Children (NAEYC) promotes high-quality early learning for all young children, birth through age 8, by connecting early childhood practice, policy, and research. We advance a diverse, dynamic early childhood profession and support all who care for, educate, and work on behalf of young children.

NAEYC members have access to award-winning publications, professional development, networking opportunities, professional liability insurance, and an array of members-only discounts.

Accreditation—NAEYC.org/accreditation

Across the country, **NAEYC Accreditation of Early Learning Programs** and **NAEYC Accreditation of Early Childhood Higher Education Programs** set the industry standards for quality in early childhood education. These systems use research-based standards to recognize excellence in the field of early childhood education.

Advocacy and Public Policy—NAEYC.org/policy

NAEYC is a leader in promoting and advocating for policies at the local, state, and federal levels that expand opportunities for all children to have equitable access to high-quality early learning. NAEYC is also dedicated to promoting policies that value early childhood educators and support their excellence.

Global Engagement—NAEYC.org/global

NAEYC's Global Engagement department works with governments and other large-scale systems to create guidelines to support early learning, as well as early childhood professionals throughout the world.

Professional Learning—NAEYC.org/ecp

NAEYC provides face-to-face training, technology-based learning, and Accreditation workshops—all leading to improvements in the knowledge, skills, and practices of early childhood professionals.

Publications and Resources —NAEYC.org/publications

NAEYC publishes some of the most valued resources for early childhood professionals, including award-winning books, *Teaching Young Children* magazine, and *Young Children*, the association's peer-reviewed journal. NAEYC publications focus on developmentally appropriate practice and enable members to stay up to date on current research and emerging trends, with information they can apply directly to their classroom practice.

Signature Events—NAEYC.org/events

NAEYC hosts three of the most important and well-attended annual events for educators, students, administrators, and advocates in the early learning community.

NAEYC's Annual Conference is the world's largest gathering of early childhood professionals.

NAEYC's Professional Learning Institute is the premier professional development conference for early childhood trainers, faculty members, researchers, systems administrators, and other professionals.

The **NAEYC Public Policy Forum** provides members with resources, training, and networking opportunities to build advocacy skills and relationships with policymakers on Capitol Hill.

NAEYC.org/membership

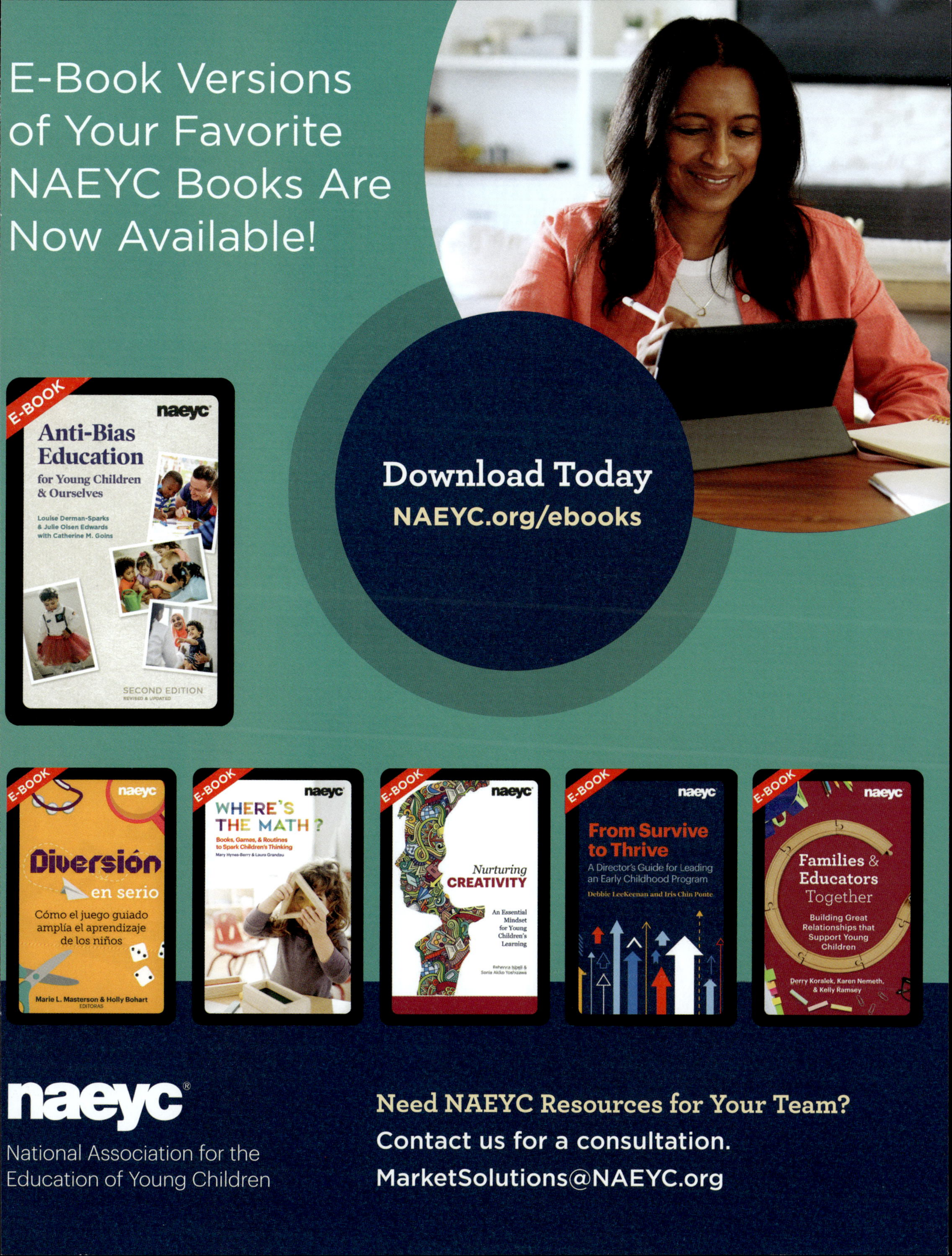
E-Book Versions of Your Favorite NAEYC Books Are Now Available!
Download Today
NAEYC.org/ebooks
E-BOOK
naeyc
Anti-Bias Education for Young Children & Ourselves
Louise Derman-Sparks & Julie Olsen Edwards with Catherine M. Goins
SECOND EDITION
E-BOOK
Diversión en serio
Cómo el juego guiado amplía el aprendizaje de los niños
Marie L. Masterson & Holly Bohart
E-BOOK
WHERE'S THE MATH?
Books, Games, & Routines to Spark Children's Thinking
E-BOOK
Nurturing CREATIVITY
An Essential Mindset for Young Children's Learning
E-BOOK
From Survive to Thrive
A Director's Guide for Leading an Early Childhood Program
Debbie LeeKeenan and Iris Chin Ponte
E-BOOK
Families & Educators Together
Building Great Relationships that Support Young Children
Derry Koralek, Karen Nemeth, & Kelly Ramsey
naeyc
National Association for the Education of Young Children
Need NAEYC Resources for Your Team?
Contact us for a consultation.
MarketSolutions@NAEYC.org

naeyc
Ethics and the Early Childhood Educator
Using the NAEYC Code
New foreword by Rhian Evans Allvin
Stephanie Feeney and Nancy K. Freeman
Third Edition
naeyc
Anti-Bias Education for Young Children & Ourselves
Louise Derman-Sparks & Julie Olsen Edwards with Catherine M. Goins
SECOND EDITION
REVISED & UPDATED
naeyc
Coaching with Powerful Interactions
A Guide for Partnering with Early Childhood Teachers
Access to 30 Videos
Judy Jablon, Amy Laura Dombro, and Shaun Johnsen
Books that Every Educator Needs!
NAEYC.org/shop
naeyc
Big Questions For Young Minds
EXTENDING CHILDREN'S THINKING
JANIS STRASSER AND LISA MUFSON BRESSON
naeyc
MAKING & TINKERING With STEM
Solving Design Challenges With Young Children
Cate Heroman